Analysis of Piezoelectric Semiconductor Structures

Jiashi Yang

Analysis of Piezoelectric Semiconductor Structures

Springer

Jiashi Yang
Lincoln, NE, USA

ISBN 978-3-030-48205-3 ISBN 978-3-030-48206-0 (eBook)
https://doi.org/10.1007/978-3-030-48206-0

This Springer imprint is published by the registered company Springer Nature Switzerland AG
The registered company address is: Gewerbestrasse 11, 6330 Cham, Switzerland

Preface

Piezoelectric materials may be dielectrics or semiconductors. In piezoelectric semiconductors, the distribution or motion of charge carriers is affected by mechanical loads through the electric field produced by piezoelectric coupling. Since the 1960s, there have been efforts on using piezoelectric semiconductors to make acoustic wave devices where the interaction between mechanical waves and electric currents is called the acoustoelectric effect. Relatively recently, various piezoelectric semiconductor structures have been synthesized such as ZnO fibers, tubes, belts, spirals, and films. They have been used to make energy harvesters for converting mechanical energy into electrical energy, transistors, and various physical as well as chemical sensors. Piezoelectric semiconductors are also used in nanostructures such as quantum wells, dots, and wires. The study of piezoelectric semiconductor materials and devices is growing rapidly. The recent studies on the topic have formed new research areas called piezotronics and piezo-phototronics.

This book is on theoretical analysis of piezoelectric semiconductor structures or devices using the phenomenological or macroscopic theory. Although most semiconductor books follow a combined microscopic and macroscopic approach, there exist purely macroscopic semiconductor models with various levels of sophistication. This book employs the simplest drift-diffusion model for semiconduction, which is electrically coupled to the macroscopic theory of piezoelectricity. This approach is relatively simple physically. Mathematically, it involves coupled differential equations, ordinary or partial. Of course the phenomenological theory employed has its limitations, but it can provide some basic understanding of certain behaviors of piezoelectric semiconductors. In addition to electromechanical couplings, thermal effects such as thermoelastic and pyroelectric couplings are also discussed. Some equations are repeated in certain sections so that they can be read independently.

Chapter 1 is a brief summary of the general three-dimensional phenomenological theory of piezoelectric semiconductors. A few exact solutions of the three-dimensional equations are presented in Chap. 2. Then the three-dimensional theory is reduced to one- and two-dimensional models for the extension of thin rods, bending of thin beams, and extension and bending of thin plates in Chaps. 3, 4, and 5, respectively, with solutions of static and dynamic problems. Chapter 6 is devoted to composite structures of piezoelectric dielectrics and nonpiezoelectric semiconductors in which the acoustoelectric effect is a consequence of the polarization and conduction in component phases. Chapters 7 specializes in thermal effects.

The literature on piezoelectric semiconductor materials and devices are abundant, with quite a few review articles available. This book is limited to theoretical results based on purely macroscopic approach. In most cases only those papers whose results are directly used in the book are listed as references. Because of the nonlinearity associated with the drift current which is the product of the electric field and carrier concentration, theoretical analysis of piezoelectric semiconductors presents considerable mathematical challenges. Most of the sections of this book are based on the linearized theory for small loads or signals. They can provide some basic understanding but are often insufficient for device design. For nonlinear analysis, numerical methods are more effective. A few numerical analyses using COMSOL are also presented in this book. As a book on piezoelectric semiconductors, the focus is on mechanically induced carrier redistribution or motion. However, some topics treated such as the effects of doping and PN junctions do not rely on electromechanical coupling and are also relevant in nonpiezoelectric semiconductors.

Lincoln, NE, USA Jiashi Yang
December, 2019

Contents

Chapter 1
Macroscopic Theory

This chapter presents a concise summary of the basic equations of the phenomenological theory of piezoelectric semiconductors. The Cartesian tensor notation is used, along with the summation convention for repeated tensor indices and the convention that a comma followed by an index denotes partial differentiation with respect to the coordinate associated with the index. A superimposed dot represents a time derivative.

1.1 Basic Theory

Piezoelectric materials may be dielectrics or semiconductors. Mechanical fields and mobile charges in piezoelectric semiconductors can interact through electric fields, which is called the acoustoelectric effect. The basic behaviors of piezoelectric semiconductors can be described by coupling the linear theory of piezoelectricity [1–4] and the macroscopic theory of nonpiezoelectric semiconductors [5, 6]. The three-dimensional phenomenological theory consists of the equation of motion (Newton's law), the charge equation of electrostatics (Gauss's law), and the conservation of charge for holes and electrons (continuity equations):

$$
\begin{aligned}
&T_{ji,j} + f_i = \rho \ddot{u}_i, \\
&D_{i,i} = q\left(p - n + N_D^+ - N_A^-\right), \\
&\dot{p} = -\frac{1}{q} J_{i,i}^p + \dot{p}\big|_{\text{thermal R-G}} + \dot{p}\big|_{\text{other processes}}, \\
&\dot{n} = \frac{1}{q} J_{i,i}^n + \dot{n}\big|_{\text{thermal R-G}} + \dot{n}\big|_{\text{other processes}},
\end{aligned}
\tag{1.1}
$$

where **T** is the stress tensor, ρ the mass density, **f** the body force which is usually zero, **u** the mechanical displacement vector, and **D** the electric displacement vector.

J. Yang, *Analysis of Piezoelectric Semiconductor Structures*,
https://doi.org/10.1007/978-3-030-48206-0_1

$q = 1.6 \times 10^{-19}$ coul is the elementary charge. p and n are the concentrations of holes and electrons. N_A^- and N_D^+ are the concentrations of ionized accepters and donors from doping. J_i^p and J_i^n are the hole and electron current densities. In the following, we will neglect those terms in Eq. $(1.1)_{3,4}$ associated with thermal recombination and generation as well as other processes of electric, magnetic, mechanical, and thermal origins. Constitutive relations accompanying Eq. (1.1) describing material behaviors can be written in the following form:

$$\begin{aligned} T_{ij} &= c_{ijkl}^E S_{kl} - e_{kij} E_k, \\ D_i &= e_{ikl} S_{kl} + \varepsilon_{ik}^S E_k, \\ J_i^p &= qp\mu_{ij}^p E_j - qD_{ij}^p p_{,j} \\ &= qp\mu_{ij}^p E_j - qpD_{ij}^p (\ln p)_{,j}, \\ J_i^n &= qn\mu_{ij}^n E_j + qD_{ij}^n n_{,j} \\ &= qn\mu_{ij}^n E_j + qnD_{ij}^n (\ln n)_{,j}, \end{aligned} \tag{1.2}$$

where $\mathbf{S}$ is the strain tensor, $\mathbf{E}$ the electric field vector, c_{ijkl}^E the elastic stiffness, e_{ijk} the piezoelectric constants, and ε_{ij}^S the dielectric constants. μ_{ij}^p and μ_{ij}^n are the carrier mobility. D_{ij}^p and D_{ij}^n are the carrier diffusion constants. $qp\mu_{ij}^p E_j$ and $qn\mu_{ij}^n E_j$ are the drift currents. $-qD_{ij}^p p_{,j}$ and $qD_{ij}^n n_{,j}$ are the diffusion currents. The piezoelectric constitutive relations in Eq. (1.2) can be written in other forms in terms of other material constants [1–4]. For example,

$$\begin{aligned} S_{ij} &= s_{ijkl}^E T_{kl} + d_{kij} E_k, \\ D_i &= d_{ikl} T_{kl} + \varepsilon_{ik}^T E_k, \end{aligned} \tag{1.3}$$

where s_{ijkl}^E are the elastic compliance, d_{ijk} the piezoelectric constants, and ε_{ij}^T the dielectric constants. The mobility and diffusion constants, e.g., μ_{33}^p, μ_{33}^n, D_{33}^p, and D_{33}^n, satisfy the Einstein relation:

$$\frac{\mu_{33}^p}{D_{33}^p} = \frac{\mu_{33}^n}{D_{33}^n} = \frac{q}{k_B T}, \tag{1.4}$$

where k_B is the Boltzmann constant and T the absolute temperature. The strain $\mathbf{S}$ and the electric field $\mathbf{E}$ are related to the mechanical displacement $\mathbf{u}$ and the electric potential φ through

$$\begin{aligned} S_{ij} &= \left(u_{i,j} + u_{j,i}\right)/2, \\ E_i &= -\varphi_{,i}. \end{aligned} \tag{1.5}$$

With successive substitutions from Eqs. (1.2) and (1.5), we can write Eq. (1.1) as six equations for **u**, φ, p, and n. This book is based on the macroscopic theory in Eqs. (1.1), (1.2), and (1.5) without the use of any information from the microscopic theory of semiconductors except Eq. (1.4) which serves as a relationship between the mobility and the diffusion constants. The right-hand side of Eq. $(1.1)_2$ is in terms of p-n. Therefore, working with p-n and p+n instead of p and n may have some mathematical advantages. $N_A^- - N_D^+$ appears as a load in Eq. $(1.1)_2$. When $N_A^- - N_D^+$ and **f** are zero, Eq. (1.1) becomes homogeneous. Then the state with **u**, φ, p, and n all being zero satisfies Eq. (1.1), which may be used as the reference state. For a finite piezoelectric semiconductor body, the electric field in the surrounding free space is usually neglected as an approximation. Then, on the surface of a finite body, the mechanical displacement vector or the traction vector, the electric potential or the normal component of **D**, and the carrier concentrations or the normal components of the current density vectors may be prescribed as boundary conditions. Integrating Eq. $(1.1)_2$ over a volume V with boundary surface S and unit outward normal **n**, we obtain

$$\int_S n_i D_i dS = \int_V q(p - n + N_D^+ - N_A^-) dV, \tag{1.6}$$

which is a relationship among the boundary normal value of **D** and doping as well as carrier concentrations. Similar relationships can also be derived from the stress equation of motion and the continuity equations, in particular their static form.

1.2 Linearization

Because of the nonlinearity in Eq. $(1.2)_{3,4}$ associated with the drift currents of holes and electrons which are the products of the unknown carrier concentrations and the unknown electric field, theoretical analyses of piezoelectric semiconductor devices may present considerable mathematical challenges. For many purposes, a linearized version of Eq. $(1.2)_{3,4}$ is sufficient. Consider the following change of unknown variables from p and n to Δp and Δn:

$$\begin{aligned} p &= p_0 + \Delta p, \\ n &= n_0 + \Delta n, \end{aligned} \tag{1.7}$$

where, for simplicity, we have denoted

$$p_0 = N_A^-, \quad n_0 = N_D^+. \tag{1.8}$$

Equation (1.7) makes the right-hand side of Eq. $(1.1)_2$ homogeneous, while at the same time makes Eq. $(1.2)_{3,4}$ nonhomogeneous. Consider the case of small Δp and Δn. We linearize Eq. $(1.2)_{3,4}$ as

$$\begin{aligned} J_i^p &\cong q p_0 \mu_{ij}^p E_j - q D_{ij}^p (p_0 + \Delta p)_{,j}, \\ J_i^n &\cong q n_0 \mu_{ij}^n E_j + q D_{ij}^n (n_0 + \Delta n)_{,j}. \end{aligned} \tag{1.9}$$

When p_0 and n_0 are uniform, Eq. (1.9) becomes homogeneous in $\mathbf{E}$, Δp, and Δn. With Δp and Δn, Eq. $(1.1)_{2\text{–}4}$ become

$$\begin{aligned} D_{i,i} &= q(\Delta p - \Delta n), \\ q\frac{\partial(\Delta p)}{\partial t} &= -J_{i,i}^p, \\ q\frac{\partial(\Delta n)}{\partial t} &= J_{i,i}^n. \end{aligned} \tag{1.10}$$

Equation $(1.10)_1$ also becomes homogeneous. Phenomenological theories involving electromechanical couplings and mechanical nonlinearities due to large deformations in semiconductors can be found in [7, 8].

1.3 One-Dimensional Problems of Crystals of Class (6mm)

We are mainly interested in ZnO which belongs to the crystal class of (6mm). Under the compact matrix notation [1–4] with indices p and q ranging from 1 to 6, when the c-axis of the crystal is along x_3, the material tensors for ZnO can be represented by the following matrices:

$$[c_{pq}] = \begin{pmatrix} c_{11} & c_{12} & c_{13} & 0 & 0 & 0 \\ c_{21} & c_{11} & c_{13} & 0 & 0 & 0 \\ c_{31} & c_{31} & c_{33} & 0 & 0 & 0 \\ 0 & 0 & 0 & c_{44} & 0 & 0 \\ 0 & 0 & 0 & 0 & c_{44} & 0 \\ 0 & 0 & 0 & 0 & 0 & c_{66} \end{pmatrix},$$

$$[e_{ip}] = \begin{pmatrix} 0 & 0 & 0 & 0 & e_{15} & 0 \\ 0 & 0 & 0 & e_{15} & 0 & 0 \\ e_{31} & e_{31} & e_{33} & 0 & 0 & 0 \end{pmatrix},$$

$$[\varepsilon_{ij}] = \begin{pmatrix} \varepsilon_{11} & 0 & 0 \\ 0 & \varepsilon_{11} & 0 \\ 0 & 0 & \varepsilon_{33} \end{pmatrix}, \tag{1.11}$$

where $c_{66}=(c_{11}-c_{12})/2$. μ_{ij} and d_{ij} have the same structure as ε_{ij}. The matrices in Eq. (1.11) are the same as those of transversely isotropic piezoelectric materials such as polarized ceramics when the poling axis is along x_3. Equation (1.11) yields the following piezoelectric constitutive relations:

$$\begin{aligned} T_{11} &= T_1 = c_{11}S_1 + c_{12}S_2 + c_{13}S_3 - e_{31}E_3, \\ T_{22} &= T_2 = c_{12}S_1 + c_{11}S_2 + c_{13}S_3 - e_{31}E_3, \\ T_{33} &= T_3 = c_{13}S_1 + c_{13}S_2 + c_{33}S_3 - e_{33}E_3, \\ T_{23} &= T_4 = c_{44}S_4 - e_{15}E_2, \\ T_{31} &= T_5 = c_{44}S_5 - e_{15}E_1, \\ T_{12} &= T_6 = c_{66}S_6, \\ D_1 &= e_{15}S_5 + \varepsilon_{11}E_1, \\ D_2 &= e_{15}S_4 + \varepsilon_{11}E_2, \\ D_3 &= e_{31}(S_1 + S_2) + e_{33}S_3 + \varepsilon_{33}E_3. \end{aligned} \tag{1.12}$$

To see the basic acoustoelectric coupling in ZnO, consider one-dimensional longitudinal motions along the c-axis satisfying $u_1 = u_2 = 0$, $\partial/\partial x_1 = 0$, and $\partial/\partial x_2 = 0$. In this case, the nontrivial fields are

$$\begin{aligned} &u_3 = u(x_3, t), \quad \varphi = \varphi(x_3, t), \\ &\Delta p = \Delta p(x_3, t), \quad \Delta n = \Delta n(x_3, t). \end{aligned} \tag{1.13}$$

From Eq. (1.5), the relevant strain and electric field components are

$$\begin{aligned} S_{33} &= u_{3,3}, \\ E_3 &= -\varphi_{,3}. \end{aligned} \tag{1.14}$$

The relevant components of T_{ij}, D_i, J_i^p, and J_i^n are

$$\begin{aligned} T_{33} &= c_{33}S_3 - e_{33}E_3 = c_{33}u_{3,3} + e_{33}\varphi_{,3}, \\ D_3 &= e_{33}S_3 + \varepsilon_{33}E_3 = e_{33}u_{3,3} - \varepsilon_{33}\varphi_{,3}, \end{aligned} \tag{1.15}$$

$$\begin{aligned} J_3^p &= qp_0\mu_{33}^p E_3 - qD_{33}^p(\Delta p)_{,3} \\ &= -qp_0\mu_{33}^p\varphi_{,3} - qD_{33}^p(\Delta p)_{,3}, \\ J_3^n &= qn_0\mu_{33}^n E_3 + qD_{33}^n(\Delta n)_{,3} \\ &= -qn_0\mu_{33}^n\varphi_{,3} + qD_{33}^n(\Delta n)_{,3}. \end{aligned} \tag{1.16}$$

In Eq. (1.16), the linearized constitutive relations in Eq. (1.9) are used and uniform doping with constant p_0 and n_0 is assumed. With successive substitutions, the relevant ones of Eq. (1.1) take the following form:

$$\begin{aligned} c_{33}u_{3,33} + e_{33}\varphi_{,33} &= \rho\ddot{u}_3, \\ e_{33}u_{3,33} - \varepsilon_{33}\varphi_{,33} &= q(\Delta p - \Delta n), \end{aligned} \tag{1.17}$$

$$\begin{aligned} \frac{\partial}{\partial t}(\Delta p) &= p_0\mu_{33}^p\varphi_{,33} + D_{33}^p(\Delta p)_{,33}, \\ \frac{\partial}{\partial t}(\Delta n) &= -n_0\mu_{33}^n\varphi_{,33} + D_{33}^n(\Delta n)_{,33}. \end{aligned} \tag{1.18}$$

For static problems, Eqs. (1.17) and (1.18) reduce to

$$\begin{aligned} c_{33}u_{3,33} + e_{33}\varphi_{,33} &= 0, \\ e_{33}u_{3,33} - \varepsilon_{33}\varphi_{,33} &= q(\Delta p - \Delta n), \end{aligned} \tag{1.19}$$

$$\begin{aligned} 0 &= p_0\mu_{33}^p\varphi_{,33} + D_{33}^p(\Delta p)_{,33}, \\ 0 &= -n_0\mu_{33}^n\varphi_{,33} + D_{33}^n(\Delta n)_{,33}. \end{aligned} \tag{1.20}$$

Equation (1.19) can be written as

$$\begin{aligned} u_{3,33} &= -\frac{e_{33}}{c_{33}}\varphi_{,33}, \\ -\bar{\varepsilon}_{33}\varphi_{,33} &= q(\Delta p - \Delta n), \end{aligned} \tag{1.21}$$

where

$$\bar{\varepsilon}_{33} = \varepsilon_{33} + \frac{e_{33}^2}{c_{33}}. \tag{1.22}$$

$\bar{\varepsilon}_{33}$ is a piezoelectrically modified dielectric constant. From Eq. (1.20), we can write

$$\begin{aligned} (\Delta p)_{,33} &= -\frac{p_0\mu_{33}^p}{D_{33}^p}\varphi_{,33}, \\ (\Delta n)_{,33} &= -\frac{n_0\mu_{33}^n}{D_{33}^n}\varphi_{,33}. \end{aligned} \tag{1.23}$$

Subtracting the two equations in Eq. (1.23) from each other, we have

$$\varphi_{,33} = -\frac{q}{k^2\bar{\varepsilon}_{33}}(\Delta p - \Delta n)_{,33}, \tag{1.24}$$

where

$$k^2 = \left(\frac{p_0\mu_{33}^p}{D_{33}^p} + \frac{n_0\mu_{33}^n}{D_{33}^n}\right)\frac{q}{\bar{\varepsilon}_{33}}. \tag{1.25}$$

Substituting Eq. (1.24) into Eq. $(1.21)_2$, we obtain a single equation for $\Delta p - \Delta n$:

$$(\Delta p - \Delta n)_{,33} = -k^2(\Delta p - \Delta n). \tag{1.26}$$

From Eq. $(1.10)_1$, it can be seen that $\Delta p - \Delta n$ is closely related to the charge from doping and carriers. We can solve Eq. (1.26) for $\Delta p - \Delta n$, then the potential from Eq. (1.24), and then the displacement and carrier concentrations from Eqs. $(1.21)_1$ and (1.23).

1.4 Antiplane Problems of Crystals of Class (6mm)

Antiplane problems [4] are relatively simple mathematically. Consider a cross section of an infinite cylindrical body of ZnO with the c-axis along the x_3 axis which is perpendicular to the cross section, and study motions with $\partial/\partial x_3 = 0$. In this case the equations split into a plane-strain problem [4] for u_1 and u_2, which is not coupled to the electrical field and therefore is not of interest here, and an antiplane problem for the following fields:

$$\begin{aligned} &u_3 = u(x_1, x_2, t), \quad \varphi = \varphi(x_1, x_2, t), \\ &\Delta p = \Delta p(x_1, x_2, t), \quad \Delta n = \Delta n(x_1, x_2, t). \end{aligned} \tag{1.27}$$

For antiplane problems the relevant strain and electric field components are

$$\begin{Bmatrix} S_5 \\ S_4 \end{Bmatrix} = \begin{Bmatrix} 2S_{31} \\ 2S_{32} \end{Bmatrix} = \nabla u, \quad \begin{Bmatrix} E_1 \\ E_2 \end{Bmatrix} = -\nabla\varphi, \tag{1.28}$$

where ∇ is the two-dimensional gradient operator. The relevant components of T_{ij}, D_i, J_i^p, and J_i^n are

$$\begin{Bmatrix} T_5 \\ T_4 \end{Bmatrix} = \begin{Bmatrix} T_{31} \\ T_{32} \end{Bmatrix} = c\nabla u + e\nabla\varphi, \tag{1.29}$$

$$\begin{Bmatrix} D_1 \\ D_2 \end{Bmatrix} = e\nabla u - \varepsilon\nabla\varphi, \tag{1.30}$$

$$\begin{aligned} \begin{Bmatrix} J_1^p \\ J_2^p \end{Bmatrix} &= -qp_0\mu^p\nabla\varphi - qD^p\nabla(\Delta p), \\ \begin{Bmatrix} J_1^n \\ J_2^n \end{Bmatrix} &= -qn_0\mu^n\nabla\varphi + qD^n\nabla(\Delta n), \end{aligned} \tag{1.31}$$

where we have denoted $c = c_{44}$, $e = e_{15}$, $\varepsilon = \varepsilon_{11}$, $\mu^p = \mu_{11}^p$, $\mu^n = \mu_{11}^n$, $D^p = D_{11}^p$, and $D^n = D_{11}^n$. The linearized constitutive relations in Eq. (1.9) are used and uniform doping with constant p_0 and n_0 is assumed. With successive substitutions, the relevant ones of Eq. (1.1) take the following form:

$$\begin{aligned} c\nabla^2 u + e\nabla^2\varphi &= \rho\ddot{u}, \\ e\nabla^2 u - \varepsilon\nabla^2\varphi &= q(\Delta p - \Delta n), \end{aligned} \tag{1.32}$$

$$\begin{aligned} \frac{\partial}{\partial t}(\Delta p) &= p_0\mu^p\nabla^2\varphi + D^p\nabla^2(\Delta p), \\ \frac{\partial}{\partial t}(\Delta n) &= -n_0\mu^n\nabla^2\varphi + D^n\nabla^2(\Delta n), \end{aligned} \tag{1.33}$$

where $\nabla^2 = \partial^2/\partial x_1^2 + \partial^2/\partial x_2^2$ is the two-dimensional Laplacian.

For static problems, Eqs. (1.32) and (1.33) reduce to

$$\begin{aligned} c\nabla^2 u + e\nabla^2\varphi &= 0, \\ e\nabla^2 u - \varepsilon\nabla^2\varphi &= q(\Delta p - \Delta n), \end{aligned} \tag{1.34}$$

$$\begin{aligned} 0 &= p_0\mu^p\nabla^2\varphi + D^p\nabla^2(\Delta p), \\ 0 &= -n_0\mu^n\nabla^2\varphi + D^n\nabla^2(\Delta n). \end{aligned} \tag{1.35}$$

Equations (1.34) and (1.35) can be decoupled as follows. From Eq. $(1.34)_1$,

$$\nabla^2 u = -\frac{e}{c}\nabla^2\varphi. \tag{1.36}$$

Substituting Eq. (1.36) into Eq. $(1.34)_2$, we obtain

$$-\bar{\varepsilon}\nabla^2\varphi = q(\Delta p - \Delta n), \tag{1.37}$$

where

$$\bar{\varepsilon} = \varepsilon + \frac{e^2}{c}. \tag{1.38}$$

We rewrite Eq. (1.35) as

$$\begin{aligned} \nabla^2(\Delta p) &= -\frac{p_0\mu^p}{D^p}\nabla^2\varphi, \\ \nabla^2(\Delta n) &= \frac{n_0\mu^n}{D^n}\nabla^2\varphi. \end{aligned} \tag{1.39}$$

Subtracting the two equations in Eq. (1.39) from each other, we have

$$\nabla^2\varphi = -\frac{q}{\bar{\varepsilon}k^2}\nabla^2(\Delta p - \Delta n), \tag{1.40}$$

where

$$k^2 = \left(\frac{p_0\mu^p}{D^p} + \frac{n_0\mu^n}{D^n}\right)\frac{q}{\bar{\varepsilon}}. \tag{1.41}$$

The substitution of Eq. (1.40) into Eq. (1.37) yields a single equation for $\Delta p - \Delta n$:

$$\nabla^2(\Delta p - \Delta n) = k^2(\Delta p - \Delta n). \tag{1.42}$$

Once $\Delta p - \Delta n$ is obtained from Eq. (1.42), φ, u, Δp, and Δn can be determined from Eqs. (1.40), (1.36), and (1.39).

1.5 Thermal Couplings

Treatments of thermal effects in nonpiezoelectric semiconductors can be found in [9]. For a thermopiezoelectric semiconductor, the three-dimensional phenomenological theory couples the linear theory of thermopiezoelectricity [4, 10, 11] and the macroscopic theory of semiconductors. It consists of the following equation of motion, the charge equation of electrostatics, the heat equation, and the conservation of charge for holes and electrons:

$$
\begin{aligned}
T_{ji,j} &= \rho \ddot{u}_i, \\
D_{i,i} &= q\left(p - n + N_D^+ - N_A^-\right), \\
\Theta_0 \dot{\eta} &= -h_{i,i} + \gamma, \\
q\dot{p} &= -J^p_{i,i}, \\
q\dot{n} &= J^n_{i,i},
\end{aligned}
\tag{1.43}
$$

where **h** is the heat flux vector, γ the body heat source, Θ_0 a reference temperature, and η the entropy density. γ includes body heat from various origins and may have a nonlinear expression such as the Joule or Ohmic heating due to conduction. Constitutive relations accompanying Eq. (1.43) can be written in the following form:

$$
\begin{aligned}
T_{ij} &= c_{ijkl}S_{kl} - e_{kij}E_k - \lambda_{ij}\theta, \\
D_i &= e_{ikl}S_{kl} + \varepsilon_{ik}E_k + p_i\theta, \\
h_i &= -\kappa_{ij}\theta_{,j} + \kappa^E_{ij}E_j, \\
\eta &= \lambda_{kl}S_{kl} + p_k E_k + \alpha\theta, \\
J^p_i &= qp\mu^p_{ij}E_j - qD^p_{ij}p_{,j} - qpD^{Tp}_{ij}\theta_{,j} \\
&= qp\mu^p_{ij}E_j - qD^p_{ij}p_{,j} - \sigma^{Tp}_{ij}\theta_{,j}, \\
J^n_i &= qn\mu^n_{ij}E_j + qD^n_{ij}n_{,j} + qnD^{Tn}_{ij}\theta_{,j} \\
&= qn\mu^n_{ij}E_j + qD^n_{ij}n_{,j} + \sigma^{Tn}_{ij}\theta_{,j},
\end{aligned}
\tag{1.44}
$$

where $|\theta| < < \Theta_0$ is a small temperature change from Θ_0. λ_{ij} are the thermoelastic constants, p_j the pyroelectric constants, κ_{ij} the heat conduction coefficients, D^{Tp}_{ij} and D^{Tn}_{ij} the thermal diffusion constants [9], κ^E_{ij}, σ^{Tp}_{ij}, and σ^{Tn}_{ij} the thermoelectric constants [12]. D^{Tp}_{ij} and D^{Tn}_{ij} may be related to D^p_{ij} and D^n_{ij} through [9]:

$$
D^{Tp} \cong \frac{D^p}{2T}, \quad D^{Tn} \cong \frac{D^n}{2T}. \tag{1.45}
$$

The heat flux and the currents in Eq. (1.44) are further restricted by the Clausius-Duhem inequality [12]. Similar to Eq. (1.3), the constitutive relations for thermopiezoelectric materials can also be written in other forms. In addition, there are other nonlinear forms of thermoelectric constitutive relations which can be found in [13].

1.6 Magnetic Couplings

When piezomagnetic effects are present and need to be considered, the piezoelectric constitutive relations in Eq. $(1.2)_{1,2}$ can be generalized to [12]

$$\begin{aligned} T_{ij} &= c_{ijkl}S_{kl} - e_{kij}E_k - h_{kij}H_k, \\ D_i &= e_{ikl}S_{kl} + \varepsilon_{ik}E_k + \alpha_{ik}H_k, \\ B_i &= h_{ikl}S_{kl} + \alpha_{ik}E_k + \mu_{ik}H_k, \end{aligned} \tag{1.46}$$

where **B** is the magnetic flux or induction, **H** the magnetic field, h_{ijk} the piezomagnetic constants, α_{ij} the magnetoelectric constants, and μ_{ij} the magnetic permeability. **B** satisfies $\nabla \cdot \mathbf{B} = 0$. Under the well-known quasistatic approximation, $\nabla \times \mathbf{H} = 0$. Then a magnetic potential ψ can be introduced through $\mathbf{H} = -\nabla\psi$. With the presence of the electric field, magnetic field, temperature gradient, electric conduction, and heat conduction, there are many cross effects such the thermomagnetic, Hall, and Ettingshausen effects [12]. Most of them are beyond the scope of the present book.

References

1. H.F. Tiersten, *Linear Piezoelectric Plate Vibrations* (Plenum, New York, 1969)
2. A.H. Meitzler, D. Berlincourt, F.S. Welsh III, H.F. Tiersten, G.A. Coquin, A.W. Warner, *IEEE Standard on Piezoelectricity* (IEEE, New York, 1988)
3. B.A. Auld, *Acoustic Fields and Waves in Solids*, vol 1 (Wiley, New York, 1973)
4. J.S. Yang, *An Introduction to the Theory of Piezoelectricity*, 2nd edn. (Springer, New York, 2018)
5. R.F. Pierret, *Semiconductor Device Fundamentals* (Pearson, Uttar Pradesh, 1996)
6. S.M. Sze, *Physics of Semiconductor Devices* (Wiley, New York, 1981)
7. H.G. de Lorenzi, H.F. Tiersten, On the interaction of the electromagnetic field with heat conducting deformable semiconductors. J. Math. Phys. **16**, 938–957 (1975)
8. G.A. Maugin, N. Daher, Phenomenological theory of elastic semiconductors. Int. J. Eng. Sci. **24**, 703–731 (1986)
9. S. Selberherr, *Analysis and Simulation of Semiconductor Devices* (Springer, New York, 1984)
10. R.D. Mindlin, On the equations of motion of piezoelectric crystals, in *Problems of Continuum Mechanics*, ed. by J.R.M. Radok, (Society for Industrial and Applied Mathematics, Philadelphia, 1961), pp. 282–290
11. R.D. Mindlin, Equations of high frequency vibrations of thermopiezoelectric crystal plates. Int. J. Solids Struct. **10**, 625–637 (1974)
12. A.C. Eringen, G.A. Maugin, *Electrodynamics of Continua*, vol I (Springer, New York, 1990)
13. D.M. Rowe, *CRC Handbook of Thermoelectrics* (CRC Press, Boca Raton, 2016)

Chapter 2
Exact Solutions

This chapter gathers a few exact solutions satisfying the three-dimensional equations of piezoelectric semiconductors. Solutions like these are possible only in a few relatively special cases. Most of them are based on the linearized theory. Section 2.1 is simple but nontrivial. Section 2.2 is a one-dimensional problem mathematically in the sense that there is only one spatial variable. Sections 2.3, 2.4, and 2.5 are antiplane problems. Section 2.3 is axisymmetric and one-dimensional. The other two are two-dimensional.

2.1 Uniform Doping

N_A^- and N_D^+ in Eq. $(1.1)_2$ may depend on spatial variables representing nonuniform doping. In the special case of uniform doping, N_A^- and N_D^+ are constants. In this case, it can be verified that

$$\begin{aligned} &p = N_A^-, \quad n = N_D^+, \\ &\mathbf{E} = 0, \quad \mathbf{D} = 0, \\ &\mathbf{J}^p = 0, \quad \mathbf{J}^n = 0, \\ &\mathbf{S} = 0, \quad \mathbf{T} = 0, \end{aligned} \tag{2.1}$$

satisfy the static form of Eqs. (1.1) and (1.2) when there is no body force, recombination, and generation. Hence, for a finite body without surface traction, charge, and currents, Eq. (2.1) is the solution of the problem. From Eq. (1.5), the displacement field can still have an undetermined rigid-body displacement. Similarly, the electric potential can still have an arbitrary constant.

J. Yang, *Analysis of Piezoelectric Semiconductor Structures*,
https://doi.org/10.1007/978-3-030-48206-0_2

2.2 Abrupt PN Junction

Consider the interface between two piezoelectric half spaces of ZnO in Fig. 2.1 [1]. The two half spaces are doped into a p region where $p > n$, and an n region where $n > p$. Then diffusion occurs at the interface to form a PN junction. Since the doping is uniform separately in the p and n regions, respectively, it has a jump discontinuity at the interface and the corresponding PN junction is called an abrupt junction [2].

Mathematically, we need to treat the two half spaces separately and then apply boundary conditions at infinity as well as continuity conditions at the interface. The problem is one-dimensional without dependence on x_1 and x_2. The governing equations are taken from Eqs. (1.26), (1.24), (1.23), and $(1.21)_1$:

$$(\Delta p - \Delta n)_{,33} = k^2(\Delta p - \Delta n), \tag{2.2}$$

$$\varphi_{,33} = -\frac{q}{k^2\overline{\varepsilon}_{33}}(\Delta p - \Delta n)_{,33}, \tag{2.3}$$

$$\begin{aligned}(\Delta p)_{,33} &= -\frac{p_0\mu_{33}^p}{D_{33}^p}\varphi_{,33},\\ (\Delta n)_{,33} &= \frac{n_0\mu_{33}^n}{D_{33}^n}\varphi_{,33},\end{aligned} \tag{2.4}$$

$$u_{3,33} = -\frac{e_{33}}{c_{33}}\varphi_{,33}, \tag{2.5}$$

where

$$k^2 = \left(\frac{p_0\mu_{33}^p}{D_{33}^p} + \frac{n_0\mu_{33}^n}{D_{33}^n}\right)\frac{q}{\overline{\varepsilon}_{33}}, \quad \overline{\varepsilon}_{33} = \varepsilon_{33} + \frac{e_{33}^2}{c_{33}}. \tag{2.6}$$

At infinity we have the following boundary conditions:

$$\begin{aligned}&T_{33}(\pm\infty) = 0, \quad D_3(\pm\infty) = 0,\\ &J_3^p(\pm\infty) = 0, \quad J_3^n(\pm\infty) = 0.\end{aligned} \tag{2.7}$$

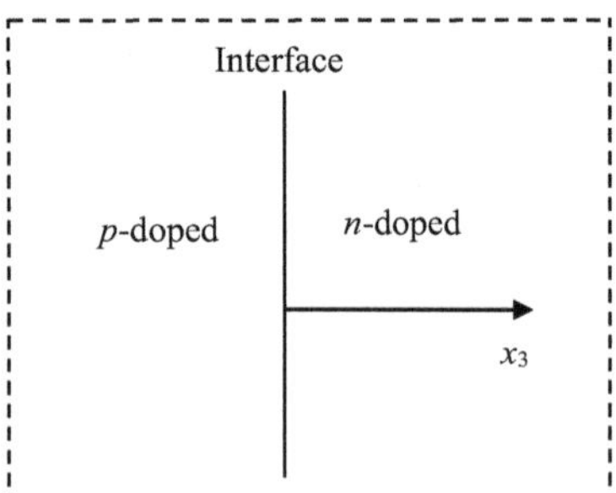

Fig. 2.1 An interface between two piezoelectric semiconductor half spaces

At the interface between the *p*- and *n*-doped regions, we have the following continuity conditions:

$$\begin{aligned} u_3(0^-) &= u_3(0^+), \quad T_{33}(0^-) = T_{33}(0^+), \\ \varphi(0^-) &= \varphi(0^+), \quad D_3(0^-) = D_3(0^+), \end{aligned} \tag{2.8}$$

$$\begin{aligned} n(0^-) &= n(0^+), \quad p(0^-) = p(0^+), \\ J_3^n(0^-) &= J_3^n(0^+), \quad J_3^p(0^-) = J_3^p(0^+). \end{aligned} \tag{2.9}$$

The above equations and boundary/continuity conditions are invariant under a rigid-body translation in the x_3 direction and a shift of the electric potential through a constant. To fix the rigid-body displacement and the arbitrary constant in the electric potential so that the displacement and potential fields are unique, we choose the interface as a reference for the displacement and potential and impose

$$u_3(0) = 0, \quad \varphi(0) = 0. \tag{2.10}$$

We have the following additional conditions representing the global conservation of holes and electrons:

$$\begin{aligned} \int_{-\infty}^{0} \Delta p dx_3 + \int_{0}^{\infty} \Delta p dx_3 = 0, \\ \int_{-\infty}^{0} \Delta n dx_3 + \int_{0}^{\infty} \Delta n dx_3 = 0. \end{aligned} \tag{2.11}$$

We use a prime for the material parameters in the *p*-doped left half space and a double prime for those in the *n*-doped region on the right.

Equations (2.2), (2.3), (2.4), and (2.5) are linear ordinary differential equations with constant coefficients. Their general solution can be obtained in a straightforward and systematic manner. For $x_3 < 0$,

$$\Delta p - \Delta n = C_1 \exp\ (k' x_3), \tag{2.12}$$

$$\varphi = -\frac{q}{(k')^2 \bar{\varepsilon}'_{33}} C_1 \exp(k' x_3) + C_2 x_3 + C_3, \tag{2.13}$$

$$u_3 = -\frac{e'_{33}}{c'_{33}} \left[-\frac{q}{(k')^2 \bar{\varepsilon}'_{33}} C_1 \exp(k' x_3) + C_2 x_3 + C_3 \right] + C_4 x_3 + C_5, \tag{2.14}$$

$$\Delta n = \frac{n'_0 \mu'^n_{33}}{D'^n_{33}} \left[-\frac{q}{(k')^2 \bar{\varepsilon}'_{33}} C_1 \exp(k' x_3) + C_2 x_3 + C_3 \right] + C_6 x_3 + C_7, \tag{2.15}$$

where C_1 through C_7 are undetermined constants. Similarly, for $x_3 > 0$,

Fig. 2.2 Doping profile

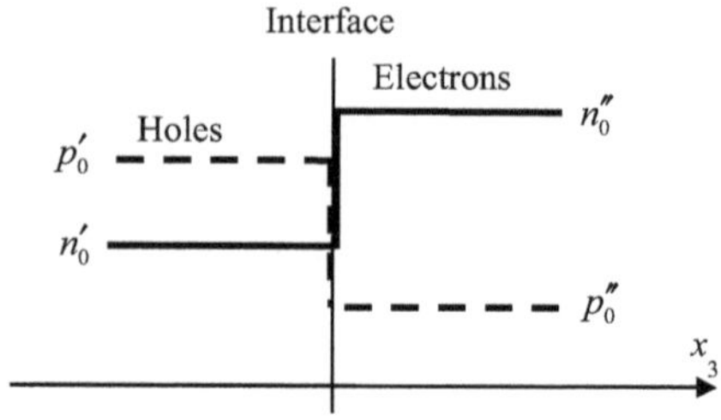

$$\Delta p - \Delta n = C_8 \exp\left(-k'' x_3\right), \tag{2.16}$$

$$\varphi = -\frac{q}{(k'')^2 \overline{\varepsilon}''_{33}} C_8 \exp\left(-k'' x_3\right) + C_9 x_3 + C_{10}, \tag{2.17}$$

$$u_3 = -\frac{e''_{33}}{c''_{33}}\left[-\frac{q}{(k'')^2 \overline{\varepsilon}''_{33}} C_8 \exp\left(-k'' x_3\right) + C_9 x_3 + C_{10}\right] + C_{11} x_3 + C_{12}, \tag{2.18}$$

$$\Delta n = \frac{n''_0 \mu''^n_{33}}{D''^n_{33}}\left[-\frac{q}{(k'')^2 \overline{\varepsilon}''_{33}} C_8 \exp\left(-k'' x_3\right) + C_9 x_3 + C_{10}\right] + C_{13} x_3 + C_{14}. \tag{2.19}$$

C_1 through C_{14} are determined by Eqs. (2.7), (2.8), (2.9), (2.10), and (2.11). Some of them can be determined in a simple manner. At the end eight constants are determined by solving eight linear algebraic equations on a computer using MATLAB.

As a numerical example consider the doping profile shown in Fig. 2.2 described by step functions. Specifically, we limit ourselves to the following doping profile with some symmetry or antisymmetry:

$$\begin{aligned} p'_0 &= 10 \times 10^{20}\ \mathrm{m}^{-3},\ p''_0 = 7 \times 10^{20}\ \mathrm{m}^{-3} = n'_0, \\ n'_0 &= 7 \times 10^{20}\ \mathrm{m}^{-3},\ n''_0 = 10 \times 10^{20}\ \mathrm{m}^{-3} = p'_0. \end{aligned} \tag{2.20}$$

The jumps of the step functions are not large so that the assumption of the linear theory is not violated. When Eq. (2.20) is true, $k' = k'' = k$. It can be seen from the above expressions that k is an important combination of parameters in piezoelectric semiconductors. For the parameters in Eq. (2.20), let

$$k_0^2 = \left(\frac{p_0 \mu^p_{33}}{D^p_{33}} + \frac{n_0 \mu^n_{33}}{D^n_{33}}\right)\frac{q}{\overline{\varepsilon}_{33}} = (p_0 + n_0)\frac{q}{k_B T}\frac{q}{\overline{\varepsilon}_{33}}, \tag{2.21}$$

which will be used as a unit for k.

Figure 2.3 shows the concentrations of holes and electrons which were initially determined by the step functions in Fig. 2.2 but are now continuous because of diffusion. They are smooth functions after diffusion although they still change rapidly near the interface. When k varies, effectively, $p_0 + n_0$ varies when other parameters are fixed. From the expressions of the fields in the above, we expect that

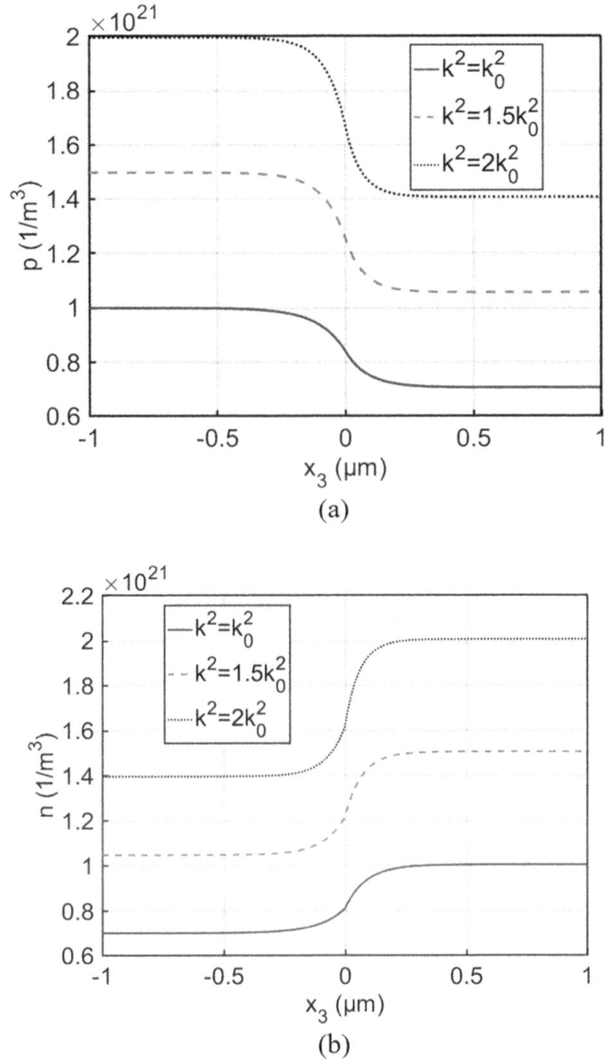

Fig. 2.3 Carrier concentrations near the interface. (**a**) $p(x_3)$ for holes. (**b**) $n(x_3)$ for electrons

when k increases, the fields vary more rapidly near the interface. This can be seen clearly in the figure.

Figure 2.4 shows effectively the distribution of the net doping and mobile charge concentration near the interface, i.e., the right-hand side of Eq. $(1.10)_1$ divided by q. This does not include the effective polarization charge. The distribution in the figure localizes near the interface and has a sign change there, showing the formation of a PN junction. These charges are often approximated by locally piecewise uniform charge distributions under the depletion layer approximation in common

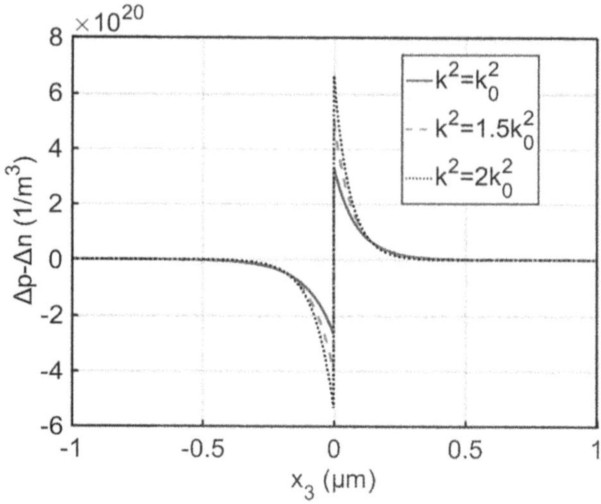

Fig. 2.4 $\Delta p - \Delta n$ near the interface

semiconductor textbooks. They can also be approximated by two oppositely charged point charges at some distance apart, forming a dipole or double layer.

Figure 2.5 shows the electric field and potential near the interface, i.e., the so-called built-in electric field and potential in a PN junction. These are familiar results in semiconductor textbooks.

Since the material is piezoelectric, the electric field causes mechanical deformation. The strain field associated with the PN junction is shown in Fig. 2.6 which localizes near the interface. In the present analysis, $T_{33} = 0$. A small T_{33} can be included to study its effects on the PN junction without mathematical difficulty. However, the above linear analysis cannot describe the typical nonlinear current-voltage relations or *I*-*V* curve of a PN junction. They will be treated by numerical methods later.

Interfaces between metal conductors and semiconductors (MS junctions) are also common and are fundamentally important in devices. The analysis of an MS junction is similar to the above and is simpler in the sense that on the metal side there is a concentrated point charge described by a delta function [2] instead of a distribution.

2.3 Circular PN Junction

Figure 2.7 shows the cross section of a circular piezoelectric semiconductor cylinder with radius *a* embedded in another unbounded piezoelectric semiconductor material. The two materials are doped oppositely and form a PN junction at their interface [3].

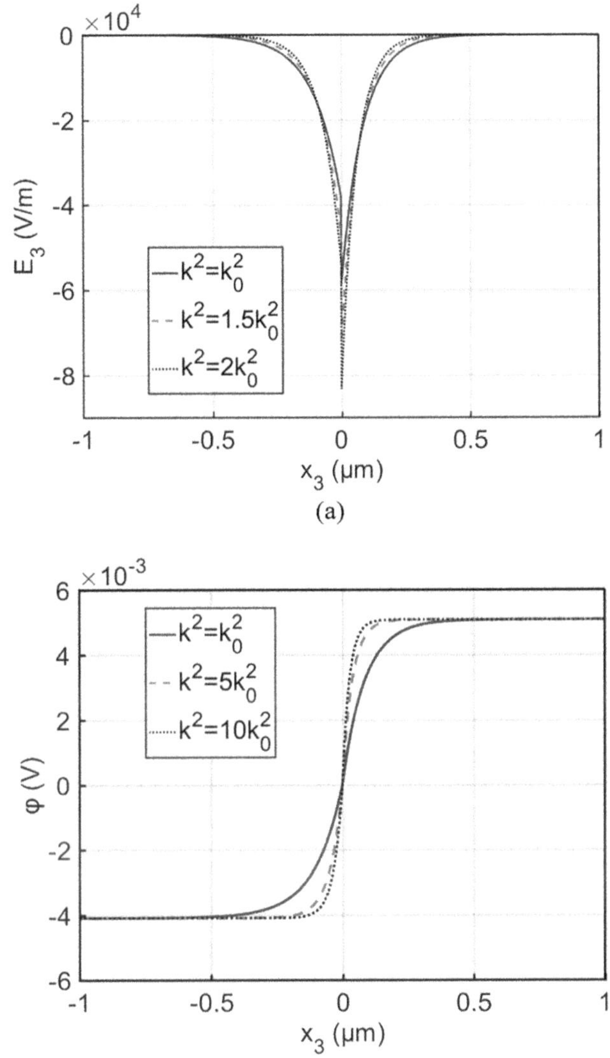

(b)

Fig. 2.5 (**a**) Electric field. (**b**) Electric Potential

Consider ZnO with the c-axis along x_3. For antiplane problems, the governing equations are from Eqs. (1.42), (1.37), (1.36), and (1.39):

$$\nabla^2(\Delta p - \Delta n) = k^2(\Delta p - \Delta n), \tag{2.22}$$

$$-\bar{\varepsilon}\nabla^2\varphi = q(\Delta p - \Delta n) \tag{2.23}$$

Fig. 2.6 Axial strain $S_3(x_3)$

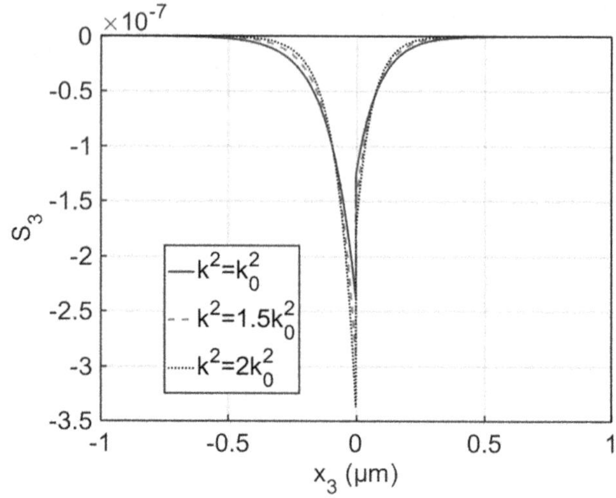

Fig. 2.7 Two piezoelectric semiconductors and their circular interface

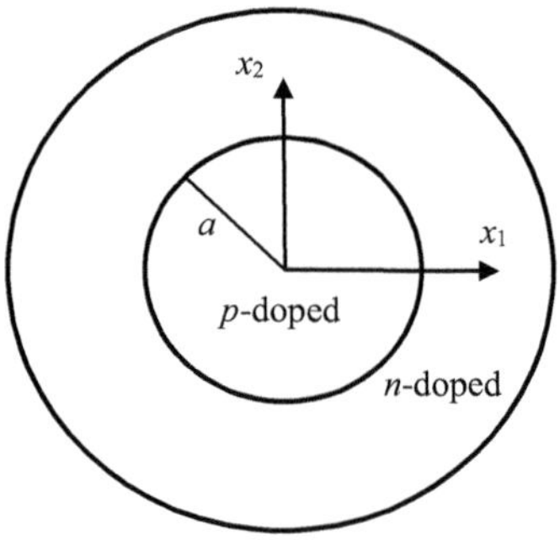

$$\nabla^2 u = -\frac{e}{c}\nabla^2\varphi, \tag{2.24}$$

$$\begin{aligned}\nabla^2(\Delta p) &= -\frac{p_0\mu^p}{D^p}\nabla^2\varphi,\\ \nabla^2(\Delta n) &= \frac{n_0\mu^n}{D^n}\nabla^2\varphi,\end{aligned} \tag{2.25}$$

where

$$k^2 = \left(\frac{p_0\mu^p}{D^p} + \frac{n_0\mu^n}{D^n}\right)\frac{q}{\varepsilon}, \tag{2.26}$$

$$\bar{\varepsilon} = \varepsilon + \frac{e^2}{c}. \tag{2.27}$$

In polar coordinates,

$$\nabla^2 = \frac{\partial^2}{\partial r^2} + \frac{1}{r}\frac{\partial}{\partial r} + \frac{1}{r^2}\frac{\partial^2}{\partial \theta^2}. \tag{2.28}$$

At the interface between the p- and n-doped regions, we have the following continuity conditions:

$$\begin{aligned} u(a^-) &= u(a^+), \quad T_{3r}(a^-) = T_{3r}(a^+), \\ \varphi(a^-) &= \varphi(a^+), \quad D_r(a^-) = D_r(a^+), \end{aligned} \tag{2.29}$$

$$\begin{aligned} n(a^-) &= n(a^+), \quad p(a^-) = p(a^+), \\ J_r^n(a^-) &= J_r^n(a^+), \quad J_r^p(a^-) = J_r^p(a^+). \end{aligned} \tag{2.30}$$

The above equations and boundary/continuity conditions are invariant under a rigid-body displacement in the x_3 direction and a shift of the electric potential through a constant. To fix the rigid-body displacement and the arbitrary constant in the electric potential so that the displacement and potential fields are unique, we choose the interface as a reference and impose

$$u(a) = 0, \quad \varphi(a) = 0. \tag{2.31}$$

We also have the following additional conditions representing the global conservation of holes and electrons:

$$\begin{aligned} \int_0^a \Delta p dr + \int_a^\infty \Delta p dr = 0, \\ \int_0^a \Delta n dr + \int_a^\infty \Delta n dr = 0. \end{aligned} \tag{2.32}$$

In addition, the fields should be finite at $r = 0$. There are restrictions on the fields at infinity too.

We use a prime for the material parameters in the p-doped central region and a double prime for the material parameters in the n-doped outer region. For the central region, the solution finite at the origin is given by

$$\Delta p - \Delta n = C_1 I_0(k' r), \tag{2.33}$$

$$\varphi = -\frac{q}{(k')^2\overline{\varepsilon}'} C_1 I_0(k'r) + C_2, \tag{2.34}$$

$$u = -\frac{e'}{c'}\left[-\frac{q}{(k')^2\overline{\varepsilon}'} C_1 I_0(k'r) + C_2\right] + C_3, \tag{2.35}$$

$$\Delta n = \frac{n'_0 \mu'^n}{D'^n}\left[-\frac{q}{(k')^2\overline{\varepsilon}'} C_1 I_0(k'r) + C_2\right] + C_4, \tag{2.36}$$

where I_0 is the zero-order modified Bessel function of the first kind. C_1 through C_4 are undetermined constants. Similarly, for $r > a$, we have

$$\Delta p - \Delta n = C_5 K_0(k''r), \tag{2.37}$$

$$\varphi = -\frac{q}{(k''_R)^2\overline{\varepsilon}''} C_5 K_0(k''r) + C_6 \ln r + C_7, \tag{2.38}$$

$$u = -\frac{e''}{c''}\left[-\frac{q}{(k''_R)^2\overline{\varepsilon}''} C_5 K_0(k''r) + C_6 \ln r + C_7\right] + C_8 \ln r + C_9, \tag{2.39}$$

$$\Delta n = \frac{n''_0 \mu''^n}{D''^n}\left[-\frac{q}{(k''_R)^2\overline{\varepsilon}''} C_5 K_0(k''r) + C_6 \ln r + C_7\right] + C_{10} \ln r + C_{11}, \tag{2.40}$$

where K_0 is the zero-order modified Bessel function of the second kind. It vanishes at infinity. The following properties of I_0 and K_0 are needed in the numerical calculation:

$$\frac{dI_0(\xi)}{d\xi} = I_1(\xi), \quad \frac{dK_0(\xi)}{d\xi} = -K_1(\xi). \tag{2.41}$$

C_1 through C_{11} are determined by Eqs. (2.29), (2.30), (2.31), and (2.32). Some of them can be determined in a simple manner. At the end eight constants are determined by solving eight linear algebraic equations on a computer using MATLAB.

As a numerical example, consider the doping profile shown in Fig. 2.8 with

$$\begin{aligned} p'_0 &= 10 \times 10^{20}\ \mathrm{m}^{-3},\ p''_0 = 7 \times 10^{20}\ \mathrm{m}^{-3} = n'_0, \\ n'_0 &= 7 \times 10^{20}\ \mathrm{m}^{-3},\ n''_0 = 10 \times 10^{20}\ \mathrm{m}^{-3} = p'_0. \end{aligned} \tag{2.42}$$

Under Eq. (2.42), we have $k' = k'' = k$. We introduce

$$k_0^2 = \left(\frac{p_0 \mu_{33}^p}{D_{33}^p} + \frac{n_0 \mu_{33}^n}{D_{33}^n}\right)\frac{q}{\overline{\varepsilon}_{33}} = (p_0 + n_0)\frac{q}{k_B T}\frac{q}{\overline{\varepsilon}_{33}}. \tag{2.43}$$

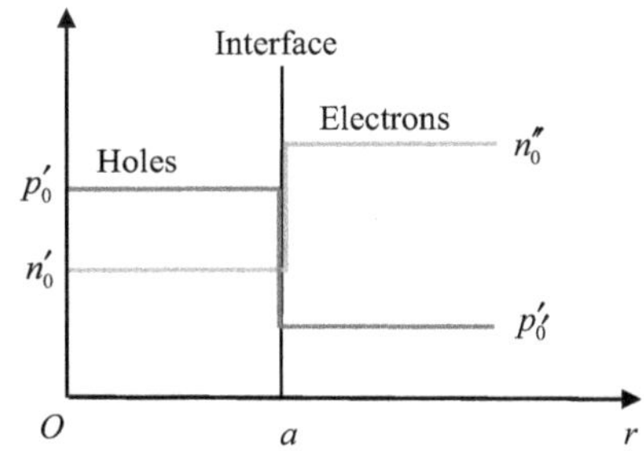

Fig. 2.8 Doping profile

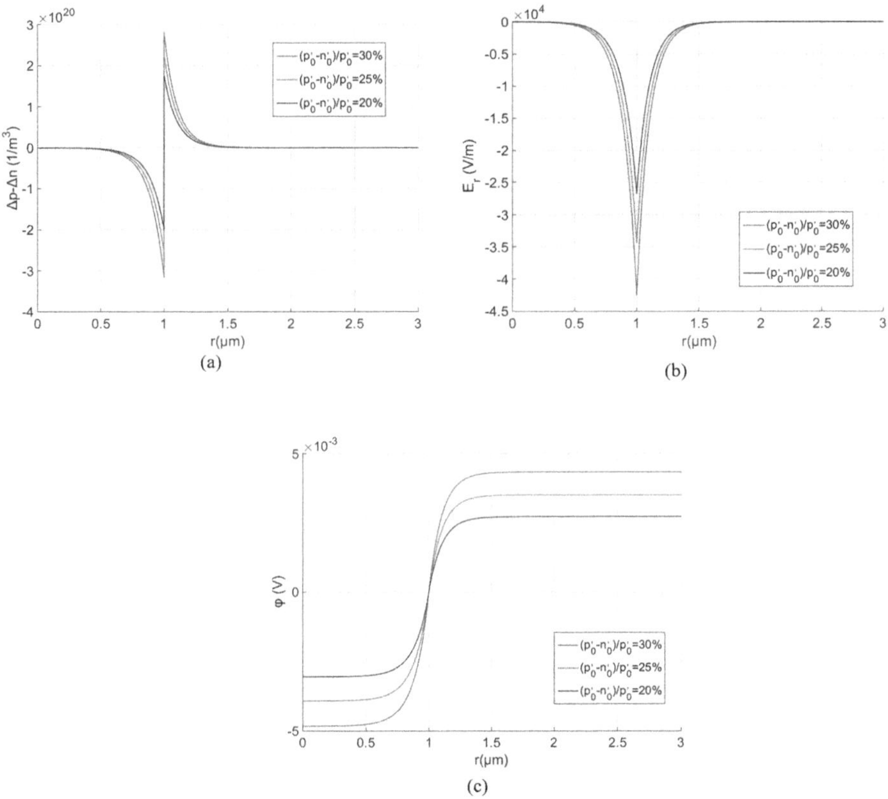

Fig. 2.9 Effects of $p_0' - n_0'$. (**a**) $\Delta p - \Delta n$ showing the formation of a PN junction. (**b**) Built-in electric field E_r. (**c**) Built-in electric potential $\varphi(r)$

Figure 2.9 shows the effects of $p_0' - n_0'$, while $k = k_0$ is kept constant. When $p_0' - n_0'$ increases, the fields in the figure become stronger and vary more rapidly as expected.

2.4 Antiplane Crack

Consider a semi-infinite crack as shown in Fig. 2.10 [4, 5]. References [4, 5] are the same. One is a minor revision of the other with some change of the title. They were both published because Professor M. Kachanov, Editor-in-Chief of *International Journal of Fracture* at the time, forwarded both versions to the publisher separately by mistake. He also said that he had made the same mistake before when contacted by the author. We limit ourselves to the case of antiplane problems of crystals of class (6mm). The material is doped into a *p*-type semiconductor. The crack surfaces are traction free and unelectroded. The electric field inside the crack is neglected. There are no free charges and normal current on the crack faces.

The governing equations are from Eqs. (1.42), (1.37), and (1.36) with $n_0 = 0$ and $\Delta n = 0$:

$$\nabla^2(\Delta p) = k^2(\Delta p), \tag{2.44}$$

$$-\overline{\varepsilon}\nabla^2\varphi = q(\Delta p), \tag{2.45}$$

$$\nabla^2 u = -\frac{e}{c}\nabla^2\varphi, \tag{2.46}$$

where

$$k^2 = \frac{p_0\mu^p}{D^p}\frac{q}{\overline{\varepsilon}}, \quad \overline{\varepsilon} = \varepsilon + \frac{e^2}{c}. \tag{2.47}$$

In polar coordinates, Eq. (2.44) takes the following form:

$$\left(\frac{\partial^2}{\partial r^2} + \frac{\partial}{r\partial r} + \frac{1}{r^2}\frac{\partial^2}{\partial\theta^2}\right)(\Delta p) - k^2(\Delta p) = 0. \tag{2.48}$$

We look for

$$\Delta p(r,\theta) = \Delta p(r)\sin\frac{\theta}{2}. \tag{2.49}$$

Substituting Eq. (2.49) into Eq. (2.48), we obtain

Fig. 2.10 A semi-infinite crack

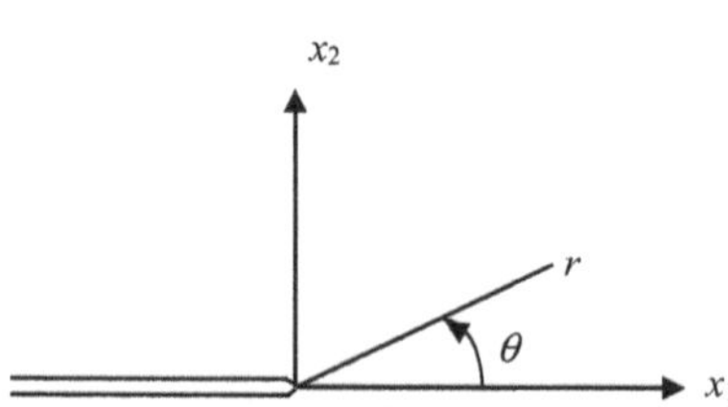

$$\frac{d}{d(kr)}\frac{d(\Delta p)}{d(kr)}+\frac{1}{kr}\frac{d(\Delta p)}{d(kr)}-\left[1+\frac{(1/2)^2}{(kr)^2}\right](\Delta p)=0. \tag{2.50}$$

Its solution is

$$\Delta p(r)=C_1I_{1/2}(kr)+C_2K_{1/2}(kr)=C_2K_{1/2}(kr), \tag{2.51}$$

where $I_{1/2}$ and $K_{1/2}$ are the first- and second-kind modified Bessel functions of order 1/2:

$$I_{1/2}(x)=\sqrt{\frac{2}{\pi x}}\sinh x,\ K_{1/2}(x)=\sqrt{\frac{2}{\pi x}}e^{-x}. \tag{2.52}$$

C_1 and C_2 are undetermined constants. Since $I_{1/2}$ is divergent for large values of its argument, we choose $C_1=0$. To find u we need to solve Eq. (2.46) which takes the following form after the use of Eqs. (2.51) and (2.45):

$$\left(\frac{\partial^2}{\partial r^2}+\frac{\partial}{r\partial r}+\frac{1}{r^2}\frac{\partial^2}{\partial\theta^2}\right)u=\frac{eq}{\bar{\varepsilon}\,c}C_2K_{1/2}(kr)\sin\frac{\theta}{2}. \tag{2.53}$$

Let

$$u(r,\theta)=u(r)\sin\frac{\theta}{2}. \tag{2.54}$$

The substitution of Eq. (2.54) into Eq. (2.53) yields

$$\left(\frac{\partial^2}{\partial r^2}+\frac{\partial}{r\partial r}-\frac{1}{4r^2}\right)u=\frac{eq}{\bar{\varepsilon}c}C_2K_{1/2}(kr). \tag{2.55}$$

The homogeneous solution of Eq. (2.55) can be obtained easily. It can be verified that its particular solution is proportional to $K_{1/2}$. Hence the general solution to Eq. (2.55) is

$$\begin{aligned}u(r,\theta)&=\left[C_3\sqrt{r}+C_4\frac{1}{\sqrt{r}}+\frac{1}{k^2}\frac{eq}{\bar{\varepsilon}c}C_2K_{1/2}(kr)\right]\sin\frac{\theta}{2}\\&=\left[C_3\sqrt{r}+C_4\frac{1}{\sqrt{r}}+\frac{1}{k^2}\frac{eq}{\bar{\varepsilon}c}C_2\sqrt{\frac{2}{\pi kr}}e^{-kr}\right]\sin\frac{\theta}{2},\end{aligned} \tag{2.56}$$

where C_3 and C_4 are undetermined constants. If we want u to be bounded at the crack tip, we must have

$$C_4 + \frac{1}{k^2}\frac{eq}{\bar{\varepsilon}c}C_2\sqrt{\frac{2}{\pi k}} = 0. \tag{2.57}$$

Then

$$u(r,\theta) = \left[C_3\sqrt{r} + C_4\frac{1}{\sqrt{r}}\left(1 - e^{-kr}\right)\right]\sin\frac{\theta}{2}. \tag{2.58}$$

Similarly, from Eq. (2.45) we have

$$\left(\frac{\partial^2}{\partial r^2} + \frac{\partial}{r\partial r} - \frac{1}{4r^2}\right)\varphi = -\frac{q}{\bar{\varepsilon}}C_2K_{1/2}(kr). \tag{2.59}$$

The relevant solution is

$$\varphi = \left[C_5\sqrt{r} + C_6\frac{1}{\sqrt{r}}\left(1 - e^{-kr}\right)\right]\sin\frac{\theta}{2}, \tag{2.60}$$

where C_6 is related to C_2 by

$$C_6 - \frac{1}{k^2}\frac{q}{\bar{\varepsilon}}C_2\sqrt{\frac{2}{\pi k}} = 0, \tag{2.61}$$

and C_5 is an undetermined constant. The strain, electric field, stress, and electric displacement components are

$$\begin{aligned} 2S_{rz} &= \frac{1}{2\sqrt{r}}\left[C_3 - C_4\frac{1}{r}\left(1 - e^{-kr}\right) + 2C_4ke^{-kr}\right]\sin\frac{\theta}{2}, \\ 2S_{\theta z} &= \frac{1}{2\sqrt{r}}\left[C_3 + C_4\frac{1}{r}\left(1 - e^{-kr}\right)\right]\cos\frac{\theta}{2}, \end{aligned} \tag{2.62}$$

$$\begin{aligned} E_r &= -\frac{1}{2\sqrt{r}}\left[C_5 - C_6\frac{1}{r}\left(1 - e^{-kr}\right) + 2C_6ke^{-kr}\right]\sin\frac{\theta}{2}, \\ E_\theta &= -\frac{1}{2\sqrt{r}}\left[C_5 + C_6\frac{1}{r}\left(1 - e^{-kr}\right)\right]\cos\frac{\theta}{2}, \end{aligned} \tag{2.63}$$

$$\begin{aligned} T_{rz} &= \frac{1}{2\sqrt{r}}\left[cC_3 + eC_5 - (cC_4 + eC_6)\frac{1}{r}\left(1 - e^{-kr}\right) + 2(cC_4 + eC_6)ke^{-kr}\right]\sin\frac{\theta}{2}, \\ T_{\theta z} &= \frac{1}{2\sqrt{r}}\left[cC_3 + eC_5 + (cC_4 + eC_6)\frac{1}{r}\left(1 - e^{-kr}\right)\right]\cos\frac{\theta}{2}, \end{aligned} \tag{2.64}$$

$$D_r = \frac{1}{2\sqrt{r}}\left[eC_3 - \varepsilon C_5 - (eC_4 - \varepsilon C_6)\frac{1}{r}\left(1 - e^{-kr}\right) + 2(eC_4 - \varepsilon C_6)ke^{-kr}\right]\sin\frac{\theta}{2},$$
$$D_\theta = \frac{1}{2\sqrt{r}}\left[eC_3 - \varepsilon C_5 + (eC_4 - \varepsilon C_6)\frac{1}{r}\left(1 - e^{-kr}\right)\right]\cos\frac{\theta}{2}. \tag{2.65}$$

It can be verified that the above solution satisfies the boundary conditions at the crack faces. When C_2, C_4, and C_6 are all equal to zero, the solution reduces to that of a piezoelectric dielectric.

2.5 Antiplane Waves in a Plate

Consider shear-horizontal (SH) waves in a plate of crystals of class (6mm) [6]. The c-axis is along x_3. The plate is unelectroded. In this problem, the electric field in the free space is considered (Fig. 2.11).

This is an antiplane problem. The governing equations for the fields in the plate are from Eqs. (1.32) and (1.33):

$$\begin{aligned} c\nabla^2 u + e\nabla^2\varphi &= \rho\ddot{u}, \\ e\nabla^2 u - \varepsilon\nabla^2\varphi &= q(\Delta p - \Delta n), \end{aligned} \tag{2.66}$$

$$\begin{aligned} \frac{\partial}{\partial t}(\Delta p) &= p_0\mu^p\nabla^2\varphi + D^p\nabla^2(\Delta p), \\ \frac{\partial}{\partial t}(\Delta n) &= -n_0\mu^n\nabla^2\varphi + D^n\nabla^2(\Delta n). \end{aligned} \tag{2.67}$$

In the free space above and below the plate, the electric potential is governed by

$$\begin{aligned} &\nabla^2\varphi = 0, \ |x_2| > h, \\ &\varphi \to 0, \ x_2 \to \pm\infty. \end{aligned} \tag{2.68}$$

The free space electric displacement is given by

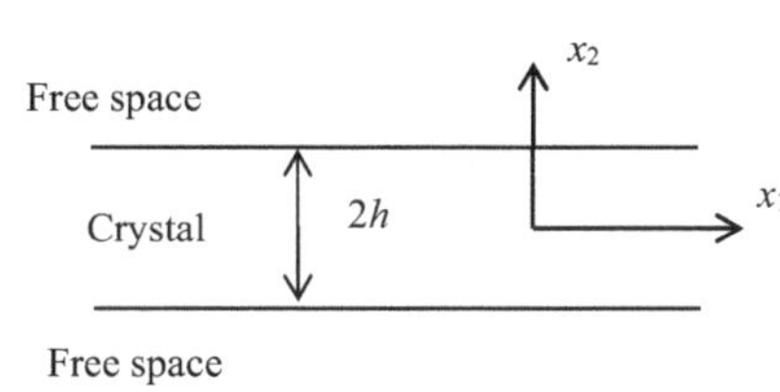

Fig. 2.11 A piezoelectric semiconductor plate and coordinate system

$$D_i = -\varepsilon_0 \varphi_{,i}. \tag{2.69}$$

At the plate surfaces, there are six boundary conditions and four continuity conditions:

$$T_{23}(\pm h) = 0, \quad J_2^p(\pm h) = 0, \quad J_2^n(\pm h) = 0, \tag{2.70}$$

$$\begin{aligned} &\varphi(h^+) = \varphi(h^-), \quad D_2(h^+) = D_2(h^-), \\ &\varphi(-h^+) = \varphi(-h^-), \quad D_2(-h^+) = D_2(-h^-). \end{aligned} \tag{2.71}$$

We look for waves propagating in the x_1 direction in the following form:

$$\begin{Bmatrix} u \\ \varphi \\ \Delta p \\ \Delta n \end{Bmatrix} = \begin{Bmatrix} A \\ B \\ C \\ D \end{Bmatrix} \exp(\eta x_2) \exp[i(\xi x_1 - \omega t)], \tag{2.72}$$

where A, B, C, and D are constants. ω is real and positive. ξ has a positive real part. The substitution of Eq. (2.72) into Eqs. (2.66) and (2.67) results in four linear homogeneous algebraic equations for A, B, C, and D. For nontrivial solutions the determinant of the coefficient matrix of the equations has to vanish, which leads to a polynomial equation of degree eight for η. Let the eight roots of this equation be $\eta^{(m)}(\xi, \omega)$ and the corresponding nontrivial solution of A, B, C, and D be $A^{(m)}$, $B^{(m)}$, $C^{(m)}$, and $D^{(m)}$. Only the ratios among $A^{(m)}$, $B^{(m)}$, $C^{(m)}$, and $D^{(m)}$ can be determined. Then the general solution to Eqs. (2.66) and (2.67) can be written as

$$\begin{Bmatrix} u \\ \varphi \\ \Delta p \\ \Delta n \end{Bmatrix} = \sum_{m=1}^{8} F^{(m)} \begin{Bmatrix} A^{(m)} \\ B^{(m)} \\ C^{(m)} \\ D^{(m)} \end{Bmatrix} \exp\left(\eta^{(m)} x_2\right) \exp[i(\xi x_1 - \omega t)], \tag{2.73}$$

where $F^{(m)}$ are undetermined constants. The solution in the free space is

$$\varphi = \begin{cases} G \exp[\xi(h - x_2)] \exp[i(\xi x_1 - \omega t)], \ x_2 > h, \\ H \exp[\xi(h + x_2)] \exp[i(\xi x_1 - \omega t)], \ x_2 < -h, \end{cases} \tag{2.74}$$

where G and H are undetermined constants. Equation (2.74) satisfies Eq. (2.68). Substituting Eqs. (2.73) and (2.74) into Eqs. (2.70) and (2.71), we obtain ten linear algebraic equations for $F^{(m)}$, G and H. For nontrivial solutions the coefficient matrix of these equations has to vanish, which yields an equation that determines the dispersion relations of ω versus ξ. The numerical procedure is carried out on a computer using MATLAB.

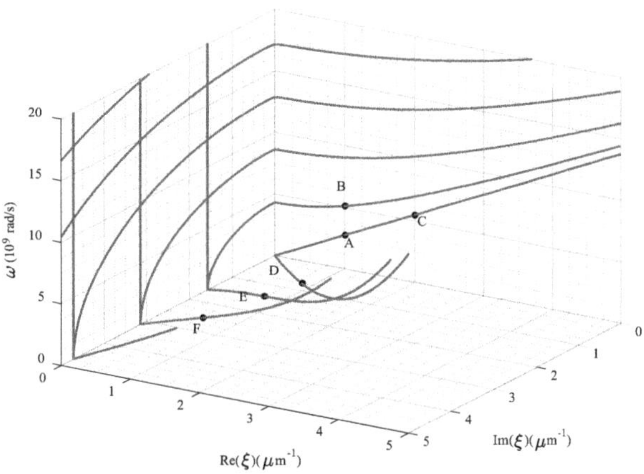

Fig. 2.12 Dispersion curves of antiplane waves

For numerical results, consider a plate of ZnO with $2h = 2\ \mu\mathrm{m}$, $p_0 = 2 \times 10^{24}\mathrm{m}^{-3}$ and $n_0 = 1.2 \times 10^{23}\mathrm{m}^{-3}$. Typical dispersion curves found numerically are shown in Fig. 2.12. The dispersion curves are complex because conduction is dissipative.

References

1. Y.X. Luo, C.L. Zhang, W.Q. Chen, J.S. Yang, An analysis of PN junctions in piezoelectric semiconductors. J. Appl. Phys. **122**, 204502 (2017)
2. R.F. Pierret, *Semiconductor Device Fundamentals* (Pearson, Uttar Pradesh, 1996)
3. Y.X. Luo, R.R. Cheng, C.L. Zhang, W.Q. Chen, J.S. Yang, Electromechanical fields near a circular PN junction between two piezoelectric semiconductors. Acta Mech. Solida Sin. **31**, 127–140 (2018)
4. J.S. Yang, A semi-infinite anti-plane crack in a piezoelectric semiconductor. Int. J. Fract. **130**, L169–L174 (2004)
5. J.S. Yang, An anti-plane crack in a piezoelectric semiconductor. Int. J. Fract. **136**, L27–L32 (2005)
6. F. Zhu, S.H. Ji, J.Q. Zhu, Z.H. Qian, J.S. Yang, Study on the influence of semiconductive property for the improvement of nanogenerator by wave mode approach. Nano Energy **52**, 474–484 (2018)

Chapter 3
Extension of Rods

Piezoelectric semiconductor rods or fibers of ZnO in extensional deformation are widely used in devices [1]. This chapter begins with the establishment of a one-dimensional model for the extension of thin rods, followed by a series of static and dynamic problems.

3.1 One-Dimensional Equations

Consider the piezoelectric semiconductor rod shown in Fig. 3.1. The shape of the cross section A may be arbitrary. The rod is assumed to be long and thin, i.e., its length is much larger than the characteristic dimension of the cross section. It is made from a piezoelectric semiconductor crystal of class (6mm) such as ZnO. The c-axis of the crystal is along the axis of the rod. The lateral surface of the rod is traction-free and is unelectroded. The electric field in the surrounding free space is neglected. These apply to the rest of the chapter.

We are interested in the extensional deformation of the rod which can be described approximately by a set of one-dimensional equations for the axial displacement u_3 and the axial stress $T_3 = T_{33}$. An important approximation for extension of thin rods is the relaxation of the lateral stress components, i.e., $T_1 = T_2 \cong 0$. In this case, from Eq. (1.3),

$$\begin{aligned} S_3 &= s_{33}^E T_3 + d_{33} E_3, \\ D_3 &= d_{33} T_3 + \varepsilon_{33}^T E_3. \end{aligned} \tag{3.1}$$

J. Yang, *Analysis of Piezoelectric Semiconductor Structures*,
https://doi.org/10.1007/978-3-030-48206-0_3

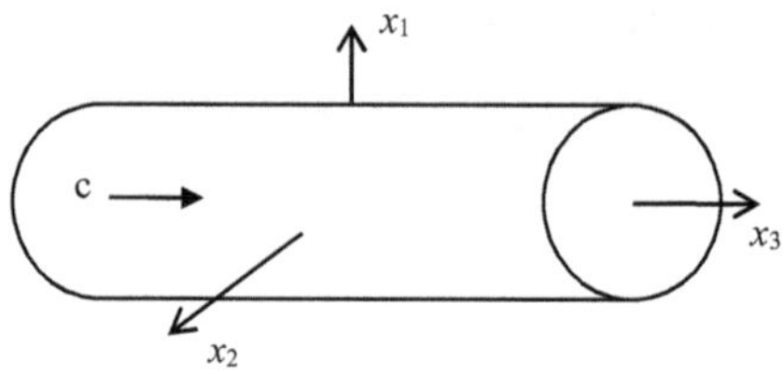

Fig. 3.1 A piezoelectric semiconductor rod of crystals of class (6mm)

We solve Eq. (3.1) for T_3 and D_3 and obtain the following one-dimensional piezoelectric constitutive relations for thin rods:

$$\begin{aligned} T_3 &= \overline{c}_{33}S_3 - \overline{e}_{33}E_3, \\ D_3 &= \overline{e}_{33}S_3 + \overline{\varepsilon}_{33}E_3, \end{aligned} \tag{3.2}$$

where $\overline{c}_{33}$, $\overline{e}_{33}$, and $\overline{\varepsilon}_{33}$ are the effective one-dimensional elastic, piezoelectric, and dielectric constants. They are related to the usual three-dimensional material constants s_{pq}^E, ε_{ij}^T, and d_{ip} through

$$\begin{aligned} \overline{c}_{33} &= 1/s_{33}^E, \\ \overline{e}_{33} &= d_{33}/s_{33}^E, \\ \overline{\varepsilon}_{33} &= \varepsilon_{33}^T - d_{33}^2/s_{33}^E. \end{aligned} \tag{3.3}$$

It can be verified that $\overline{\varepsilon}_{33} + \overline{e}_{33}^2/\overline{c}_{33} = \varepsilon_{33}^T$. We will write ε_{33}^T as ε_{33} in the rest of this chapter whenever there is no ambiguity. The nonlinear one-dimensional constitutive relations for currents are taken from Eq. $(1.2)_{3,4}$:

$$\begin{aligned} J_3^p &= qp\mu_{33}^p E_3 - qD_{33}^p \frac{\partial p}{\partial x_3}, \\ J_3^n &= qn\mu_{33}^n E_3 + qD_{33}^n \frac{\partial n}{\partial x_3}. \end{aligned} \tag{3.4}$$

The one-dimensional field equations are from Eq. (1.1):

$$\begin{aligned} &\frac{\partial T_3}{\partial x_3} + f_3 = \rho \ddot{u}_3, \\ &\frac{\partial D_3}{\partial x_3} = q\left(p - n + N_D^+ - N_A^-\right) , \\ &q\dot{p} = -\frac{\partial J_3^p}{\partial x_3}, \\ &q\dot{n} = \frac{\partial J_3^n}{\partial x_3}. \end{aligned} \tag{3.5}$$

The relevant strain-displacement relation and the electric field-potential relation are from Eq. (1.5):

$$S_3 = \frac{\partial u_3}{\partial x_3},$$
$$E_3 = -\frac{\partial \varphi}{\partial x_3}. \tag{3.6}$$

Consider the special case of a rod in static equilibrium without body force, i.e., $f_3 = 0$. In this case, from Eq. $(3.5)_1$,

$$T_3 = C_1, \tag{3.7}$$

where C_1 is an integration constant. From Eq. $(3.2)_1$,

$$S_3 = \frac{1}{\bar{c}_{33}}(C_1 + \bar{e}_{33}E_3). \tag{3.8}$$

Substituting Eq. (3.8) and Eq. $(3.6)_2$ into Eq. $(3.2)_2$, we obtain

$$D_3 = \frac{\bar{e}_{33}}{\bar{c}_{33}}C_1 + \left(\bar{\varepsilon}_{33} + \frac{\bar{e}_{33}^2}{\bar{c}_{33}}\right)E_3 = \frac{\bar{e}_{33}}{\bar{c}_{33}}C_1 - \varepsilon_{33}\varphi_{,3}. \tag{3.9}$$

For equilibrium, from Eq. $(3.5)_{3,4}$, J_3^p and J_3^n are constants. We consider the case when these constants are zero. Then

$$J_3^p = -qp\mu_{33}^p\frac{\partial \varphi}{\partial x_3} - qD_{33}^p\frac{\partial p}{\partial x_3} = 0,$$
$$J_3^n = -qn\mu_{33}^n\frac{\partial \varphi}{\partial x_3} + qD_{33}^n\frac{\partial n}{\partial x_3} = 0. \tag{3.10}$$

With the use of Eq. (1.4), we rewrite Eq. (3.10) as

$$\frac{1}{p}\frac{\partial p}{\partial x_3} = -\frac{q}{k_BT}\frac{\partial \varphi}{\partial x_3},$$
$$\frac{1}{n}\frac{\partial n}{\partial x_3} = \frac{q}{k_BT}\frac{\partial \varphi}{\partial x_3}. \tag{3.11}$$

Equation (3.11) can be integrated to produce

$$p = p^0 \exp\left(-\frac{q}{k_BT}\varphi\right),$$
$$n = n^0 \exp\left(\frac{q}{k_BT}\varphi\right), \tag{3.12}$$

where p^0 and n^0 are integration constants. Physically they are the values of p and n at $\varphi = 0$. We note that the definitions of p^0 and n^0 here in Eq. (3.12) are different from the p_0 and n_0 in Eq. (1.8). Equation (3.12) implies that $pn = p^0 n^0$ which is a constant. Within the macroscopic theory we are using, p^0 and n^0 are integration constants and can be determined from, e.g., boundary conditions. Therefore, in general, p^0 and n^0 depend on the electromechanical loads and structural parameters in addition to material properties. The substitution of Eqs. (3.9) and (3.12) into Eq. $(3.5)_2$ gives the following equation governing the electric potential:

$$\begin{aligned} &-\varepsilon_{33}\varphi_{,33} \\ &= q\left[p^0 \exp\left(-\frac{q}{k_B T}\varphi\right) - n^0 \exp\left(\frac{q}{k_B T}\varphi\right) + N_D^+(x_3) - N_A^-(x_3)\right]. \end{aligned} \tag{3.13}$$

For small φ, we make the following approximation in Eq. (3.12):

$$\begin{aligned} p &\cong p^0\left(1 - \frac{q}{k_B T}\varphi\right), \\ n &\cong n^0\left(1 + \frac{q}{k_B T}\varphi\right), \end{aligned} \tag{3.14}$$

which can describe small carrier concentration perturbations and low electric potential. Substituting Eq. (3.14) into Eq. (3.13), we obtain a linear equation for the potential:

$$\varphi_{,33} = -\frac{q}{\varepsilon_{33}}\left[p^0 - n^0 - \left(p^0 + n^0\right)\frac{q}{k_B T}\varphi + N_D^+(x_3) - N_A^-(x_3)\right], \tag{3.15}$$

which can be further written as

$$\varphi_{,33} - k^2\varphi = -\frac{q}{\varepsilon_{33}}\left[p^0 - n^0 + N_D^+(x_3) - N_A^-(x_3)\right], \tag{3.16}$$

where

$$\begin{aligned} k^2 &= \frac{q}{\varepsilon_{33}}\left(p^0 + n^0\right)\frac{q}{k_B T} = \frac{1}{\lambda_D^2}, \\ \lambda_D^2 &= \frac{\varepsilon_{33} k_B T}{(p^0 + n^0) q^2}. \end{aligned} \tag{3.17}$$

The general solution to Eq. (3.16) can be written as

$$\varphi = C_2 \cosh kx_3 + C_3 \sinh kx_3 + \frac{q}{k^2 \varepsilon_{33}}\left(p^0 - n^0\right) + \varphi^p(x_3), \tag{3.18}$$

where C_2 and C_3 are integration constants. φ^p is a particular solution of the following nonhomogeneous equation:

$$\varphi^p_{,33} - k^2\varphi^p = -\frac{q}{\varepsilon_{33}}\left[N_D^+(x_3) - N_A^-(x_3)\right]. \tag{3.19}$$

Once φ is known, p and n can be obtained from Eq. (3.14). For the mechanical displacement, from Eqs. (3.8) and $(3.6)_2$, we have, after the integration with respect to x_3 once,

$$u_3 = \frac{1}{\bar{c}_{33}}(C_1 x_3 - \bar{e}_{33}\varphi) + C_4, \tag{3.20}$$

where C_4 is an integration constant. The polarization and the effective polarization charge density can be calculated from the following well-known expressions:

$$\begin{aligned} P_3 &= D_3 - \varepsilon_0 E_3, \\ \rho^P &= -P_{k,k} = -P_{3,3}. \end{aligned} \tag{3.21}$$

In the studies of other systems of mobile charges, e.g., plasmas and electrolytes, Eq. $(3.5)_2$ is called the Poisson's equation, Eq. (3.12) the Boltzmann distribution, Eq. (3.13) the Poisson-Boltzmann equation, Eq. (3.16) the Debye-Hückel equation, and λ_D the Debye-Hückel length.

3.2 Linear Doping

Consider the simple case of linear doping described by

$$\begin{aligned} N_A^-(x_3) &= b_1 x_3 + c_1, \\ N_D^+(x_3) &= b_2 x_3 + c_2. \end{aligned} \tag{3.22}$$

Hence

$$\begin{aligned} &N_D^+(x_3) - N_A^-(x_3) = b x_3 + c, \\ &b = b_2 - b_1, \quad c = c_2 - c_1. \end{aligned} \tag{3.23}$$

In this case a particular solution of Eq. (3.19) is simply

$$\varphi^p = \frac{q}{\varepsilon_{33}k^2}(b x_3 + c). \tag{3.24}$$

Therefore, the direct contributions of linear doping as a particular solution to the electric potential and hence p as well as n according to Eq. (3.14) are also linear. The corresponding drift and diffusion currents are constants. Since the electric potential in Eq. (3.14) has to be small, the above observations are true only when $|x_3|$ is small.

3.3 Linearly Graded PN Junction

The PN junctions between two piezoelectric semiconductor half spaces and between a circular cylinder and its surrounding material treated in Sects. 2.2 and 2.3 are abrupt junctions with piecewise constant doping profiles. In reality, the doping profiles are gradually changing or graded. In this section we study linearly graded doping and PN junctions. Consider a rod with the doping profile in Fig. 3.2 [2]. The left half is mainly p-doped and the right half mainly n-doped. Between the two halves, there is a finite transition zone of width $2w$ in which the doping varies linearly. The PN junctions in Sects. 2.2 and 2.3 correspond to the special case of $w = 0$.

Quantitatively, the doping profile in Fig. 3.2 is described by

$$N_A^- = \begin{cases} a_1, & -L < x_3 < -w, \\ a_1 + \dfrac{a_2 - a_1}{2w}(x_3 + w), & |x_3| < w, \\ a_2, & w < x_3 < L, \end{cases} \tag{3.25}$$

and

$$N_D^+ = \begin{cases} b_1, & -L < x_3 < -w, \\ b_1 + \dfrac{b_2 - b_1}{2w}(x_3 + w), & |x_3| < w, \\ b_2, & w < x_3 < L. \end{cases} \tag{3.26}$$

Then,

$$N_D^+ - N_A^- = \begin{cases} d_1, & -L < x_3 < -w, \\ d_3 + d_4 x_3, & |x_3| < w, \\ d_2, & w < x_3 < L, \end{cases} \tag{3.27}$$

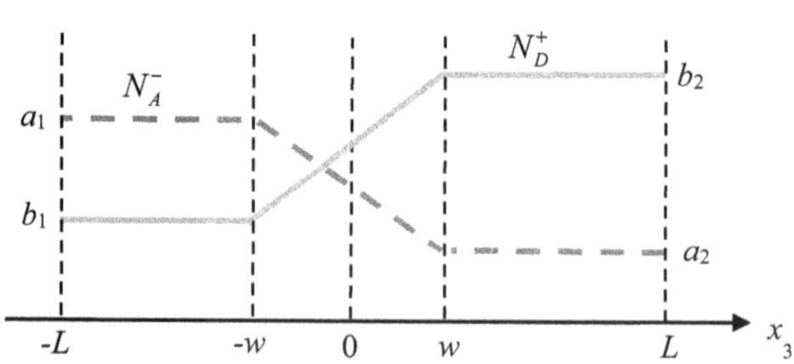

Fig. 3.2 Linearly graded doping

where

$$
\begin{aligned}
&d_1 = b_1 - a_1, \quad d_2 = b_2 - a_2, \\
&d_3 = \frac{b_1 - a_1}{2} + \frac{b_2 - a_2}{2}, \quad d_4 = \frac{b_2 - b_1}{2w} - \frac{a_2 - a_1}{2w}.
\end{aligned} \tag{3.28}
$$

We consider a free rod with the following boundary and continuity conditions:

$$
\begin{aligned}
&T_{33}(\pm L) = 0, \quad D_3(\pm L) = 0, \\
&J_3^p(\pm L) = 0, \quad J_3^n(\pm L) = 0,
\end{aligned} \tag{3.29}
$$

$$
\begin{aligned}
&u_3(\pm w^-) = u_3(\pm w^+), \quad \varphi(\pm w^-) = \varphi(\pm w^+), \\
&p(\pm w^-) = p(\pm w^+), \quad n(\pm w^-) = n(\pm w^+),
\end{aligned} \tag{3.30}
$$

$$
\begin{aligned}
&T_{33}(\pm w^-) = T_{33}(\pm w^+), \quad D_3(\pm w^-) = D_3(\pm w^+), \\
&J_3^p(\pm w^-) = J_3^p(\pm w^+), \quad J_3^n(\pm w^-) = J_3^n(\pm w^+).
\end{aligned} \tag{3.31}
$$

To make the displacement and potential unique, we impose

$$
u_3(0) = 0, \qquad \varphi(0) = 0. \tag{3.32}
$$

In addition, we also have the following global charge neutrality condition:

$$
\int_{-L}^{L} \left(p - N_A^-\right) \, dx_3 = 0. \tag{3.33}
$$

Since Eq. $(3.5)_2$ and the relevant boundary as well as continuity conditions imply that

$$
\int_{-L}^{L} \left(p - n + N_D^+ - N_A^-\right) \, dx_3 = 0, \tag{3.34}
$$

Eqs. (3.33) and (3.34) together further imply that

$$
\int_{-L}^{L} \left(-n + N_D^+\right) \, dx_3 = 0. \tag{3.35}
$$

Therefore Eq. (3.33) is the only independent charge neutrality condition.

Consider the special and relatively simple case when $L = \infty$. In this case, we have, for $x_3 < -w$,

$$\begin{aligned}
&\varphi^p = \frac{q}{\varepsilon_{33}k^2}d_1, \\
&\varphi = C_1 \exp k(x_3 + w) + \frac{q}{k^2\varepsilon_{33}}(p^0 - n^0) + \varphi^p(x_3), \\
&p \cong p^0\left(1 - \frac{q}{k_B T}\varphi\right), \quad n \cong n^0\left(1 + \frac{q}{k_B T}\varphi\right), \\
&u_3 = -\frac{\bar{e}_{33}}{\bar{c}_{33}}\varphi + C_2, \\
&D_3 = -\varepsilon_{33}\varphi_{,3}, \quad T_3 = 0, \quad J_3^p = 0, \quad J_3^n = 0,
\end{aligned} \tag{3.36}$$

and, similarly, for $x_3 > w$,

$$\begin{aligned}
&\varphi^p = \frac{q}{\varepsilon_{33}k^2}d_2, \\
&\varphi = C_3 \exp k(w - x_3) + \frac{q}{k^2\varepsilon_{33}}(p^0 - n^0) + \varphi^p(x_3), \\
&p \cong p^0\left(1 - \frac{q}{k_B T}\varphi\right), \quad n \cong n^0\left(1 + \frac{q}{k_B T}\varphi\right), \\
&u_3 = -\frac{\bar{e}_{33}}{\bar{c}_{33}}\varphi + C_4, \\
&D_3 = -\varepsilon_{33}\varphi_{,3}, \quad T_3 = 0, \quad J_3^p = 0, \quad J_3^n = 0.
\end{aligned} \tag{3.37}$$

Inside the transition zone where $|x_3| < w$, we have

$$\begin{aligned}
&\varphi^p = \frac{q}{\varepsilon_{33}k^2}(d_3 + d_4 x_3), \\
&\varphi = C_5 \sinh k(x_3 + w) + C_6 \sinh k(x_3 - w) \\
&\quad + \frac{q}{k^2\varepsilon_{33}}(p^0 - n^0) + \varphi^p(x_3), \\
&p \cong p^0\left(1 - \frac{q}{k_B T}\varphi\right), \quad n \cong n^0\left(1 + \frac{q}{k_B T}\varphi\right), \\
&u_3 = -\frac{\bar{e}_{33}}{\bar{c}_{33}}\varphi + C_7, \\
&D_3 = -\varepsilon_{33}\varphi_{,3}, \quad T_3 = 0, \quad J_3^p = 0, \quad J_3^n = 0.
\end{aligned} \tag{3.38}$$

C_1–C_7, p^0, and n^0 are determined by Eqs. (3.29), (3.30), (3.31), (3.32), and (3.33) using a computer.

As an example, consider the following case with some symmetry and/or antisymmetry:

$$\begin{aligned}
&a_1 = b_2 = 1.0 \times 10^{21}\ \text{m}^{-3}, \\
&a_2 = b_1 = 0.8 a_1, \\
&w = 0.1\ \mu\text{m}.
\end{aligned} \tag{3.39}$$

In this case, it is found that

$$p^0 = n^0 = 0.9 \times 10^{21}\ \mathrm{m}^{-3} = (a_1 + a_2)/2 = (b_1 + b_2)/2, \tag{3.40}$$

$$\frac{1}{k} = 1.005 \times 10^{-7} \cong 0.1\ \ \mu\mathrm{m}. \tag{3.41}$$

We denote the right-hand side of Eq. $(3.5)_2$, i.e., the sum of the doping and mobile charges, by

$$\rho^e = q\left(p - n + N_D^+ - N_A^-\right). \tag{3.42}$$

Figure 3.3 shows the effects of $2w$, i.e., the width of the transition zone, on ρ^e, the electric field, and the electric potential. As $2w$ decreases, the transition zone is narrower, and all fields change more rapidly there. $\rho^e$and the electric field become stronger. The built-in voltage is not sensitive to w because, as w decreases, the stronger electric field is over a narrower region, and thus the voltage as the spatial integration of the electric field does not change much. Far away from the transition zone, the fields are insensitive to w.

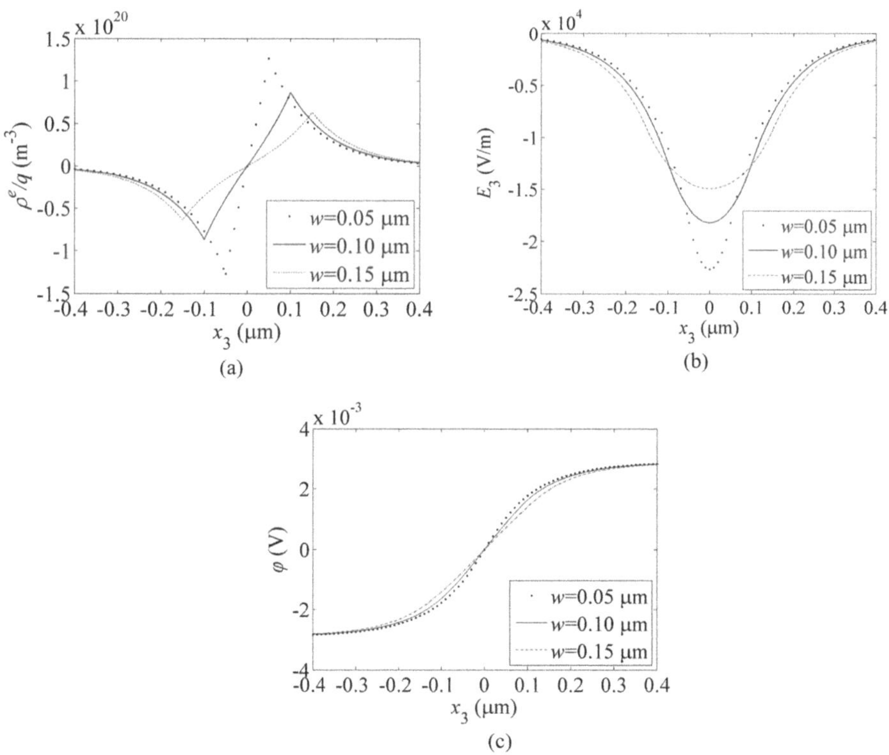

Fig. 3.3 Effects of the width of the transition zone. (**a**) ρ^e/q. (**b**) $E_3(x_3)$. (**c**) $\varphi(x_3)$

3.4 Smoothly Graded PN Junction

Consider a rod of length $2L$ with a general doping profile that varies smoothly [3]. The rod is in equilibrium without body force and currents, i.e., $f_3 = 0$, $J_3^p = 0$, and $J_3^n = 0$. In this case, from Eqs. (3.16), (3.14), and (3.20), the governing equations are

$$\varphi_{,33} - k^2\varphi = -\frac{q}{\varepsilon_{33}}\left[p^0 - n^0 + N_D^+(x_3) - N_A^-(x_3)\right], \tag{3.43}$$

$$p \cong p^0\left(1 - \frac{q}{k_BT}\varphi\right), \quad n \cong n^0\left(1 + \frac{q}{k_BT}\varphi\right), \tag{3.44}$$

$$u_3 = \frac{1}{\bar{c}_{33}}(C_1x_3 - \bar{e}_{33}\varphi) + C_4, \tag{3.45}$$

where

$$k^2 = \frac{q}{\varepsilon_{33}}\left(p^0 + n^0\right)\frac{q}{k_BT}. \tag{3.46}$$

We consider a free rod with the following boundary conditions:

$$\begin{aligned} &T_{33}(\pm L) = 0, \quad D_3(\pm L) = 0, \\ &J_3^p(\pm L) = 0, \quad J_3^n(\pm L) = 0. \end{aligned} \tag{3.47}$$

To make the displacement and the electric potential unique, we impose

$$u_3(-L) = 0, \quad \varphi(-L) = 0. \tag{3.48}$$

Since we have set $\varphi = 0$ at $x_3 = -L$, we have

$$p(-L) = p^0, \quad n(-L) = n^0. \tag{3.49}$$

We are interested in a PN junction near $x_3 = 0$. Since a PN junction produces a local charge distribution and a local electric field, when L is sufficiently large, p^0 and n^0 as defined by Eq. (3.49) can be determined from doping approximately, i.e.,

$$p^0 \cong N_A^-(-L), \quad n^0 \cong N_D^+(-L). \tag{3.50}$$

The general solution to Eq. (3.43) can be written as

$$\begin{aligned} \varphi =& C_2 \sinh k(x_3 + L) + C_3 \sinh k(x_3 - L) \\ &+ \frac{q}{k^2\varepsilon_{33}}\left(p^0 - n^0\right) + \varphi^p(x_3), \end{aligned} \tag{3.51}$$

where φ^p satisfies

$$\varphi^p_{,33} - k^2\varphi^p = -\frac{q}{\varepsilon_{33}}\left[N_D^+(x_3) - N_A^-(x_3)\right]. \tag{3.52}$$

Consider the following doping described by a power series in general:

$$N_D^+(x_3) - N_A^-(x_3) = \sum_{n=0}^{\infty} a_n x_3^n. \tag{3.53}$$

Let

$$\varphi^p = \sum_{n=0}^{\infty} c_n x_3^n. \tag{3.54}$$

Substituting Eqs. (3.53) and (3.54) into Eq. (3.52), we obtain

$$\begin{aligned}&(n+2)(n+1)c_{n+2} - k^2 c_n = -\frac{q}{\varepsilon_{33}} a_n,\\ &n = 0, 1, 2, \cdots.\end{aligned} \tag{3.55}$$

The solution to Eq. (3.55) is not unique. We choose a specific solution of Eq. (3.55) with $c_0 = 0$ and $c_1 = 0$. Then the rest of c_n can be calculated from Eq. (3.55) as a recurrence relation.

As an example of a PN Junction, we consider the specific doping described by the error function below:

$$\begin{aligned}N_A^-(x_3) &= \alpha_1 + \beta_1 \mathrm{erf}(\lambda x_3),\\ N_D^+(x_3) &= \alpha_2 + \beta_2 \mathrm{erf}(\lambda x_3).\end{aligned} \tag{3.56}$$

Then

$$\begin{aligned}&N_D^+(x_3) - N_A^-(x_3) = a + b\mathrm{erf}(\lambda x_3),\\ &a = \alpha_2 - \alpha_1, \quad b = \beta_2 - \beta_1.\end{aligned} \tag{3.57}$$

Since

$$\mathrm{erf}(\lambda x_3) = \frac{2}{\sqrt{\pi}} \sum_{n=0}^{\infty} (-1)^n \frac{\lambda^{2n+1}}{n!(2n+1)} x_3^{2n+1}, \tag{3.58}$$

we have, from Eqs. (3.53) and $(3.57)_1$,

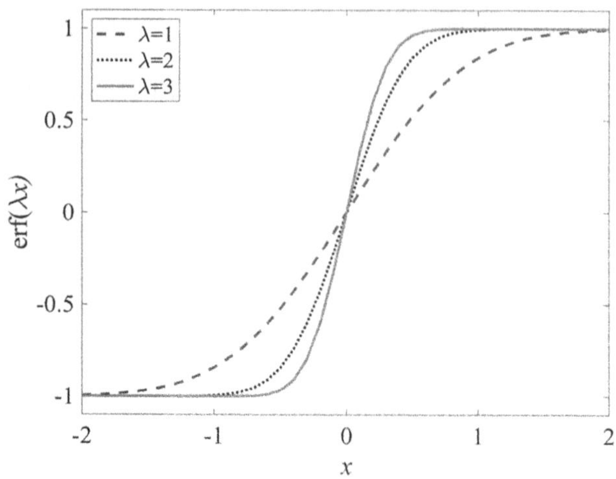

Fig. 3.4 Basic behavior of error function

$$\sum_{n=0}^{\infty} a_n x_3^n = a + b\frac{2}{\sqrt{\pi}}\sum_{n=0}^{\infty}(-1)^n\frac{\lambda^{2n+1}}{n!(2n+1)}x_3^{2n+1}. \quad (3.59)$$

We plot $\mathrm{erf}(\lambda x_3)$ in Fig. 3.4 which converges everywhere. Its behavior is suitable for describing the doping profiles of certain PN junctions with symmetry or antisymmetry. $\tan^{-1}(\lambda x_3)$ and $\tanh(\lambda x_3)$ have similar behaviors but are with a finite domain of convergence.

As a numerical example, consider a rod with

$$\begin{aligned} &L = 6\,\mu\mathrm{m}, \quad \lambda = 10^7\,\mathrm{m}^{-1}, \\ &\alpha_1 = 0.9\times 10^{21}\,\mathrm{m}^{-3}, \quad \beta_1 = -0.1\times 10^{21}\,\mathrm{m}^{-3}, \\ &\alpha_2 = 0.9\times 10^{21}\,\mathrm{m}^{-3}, \quad \beta_2 = 0.1\times 10^{21}\,\mathrm{m}^{-3}. \end{aligned} \quad (3.60)$$

In this case $a = 0$ according to Eq. (3.57) and $N_D^+ - N_A^- = b\,\mathrm{erf}(\lambda x_3)$ is an odd function. Instead of Eqs. (3.48), (3.49), and (3.50), to make use of the symmetry or antisymmetry of the doping profile implied by Eq. (3.60), we simply impose

$$\begin{aligned} &u_3(0) = 0, \quad \varphi(0) = 0, \\ &p^0 = N_A^-(0) = \alpha_1, \\ &n^0 = N_D^+(0) = \alpha_2 = \alpha_1. \end{aligned} \quad (3.61)$$

In Fig. 3.5, (a)–(c) show the effects of λ on the normalized doping and mobile charge ρ^e/q whose definition is in Eq. (3.42), the effective polarization charge ρ^P/q given by Eq. (3.21), the total charge $(\rho^e + \rho^P)/q$, the electric field, and the electric potential. ρ^e and ρ^P are with opposite signs, and ρ^e dominates. In addition to the distributed ρ^P, for a finite rod, in general, there may also exist concentrated effective polarization charges at the two ends of the rod. These concentrated charges depend on the values

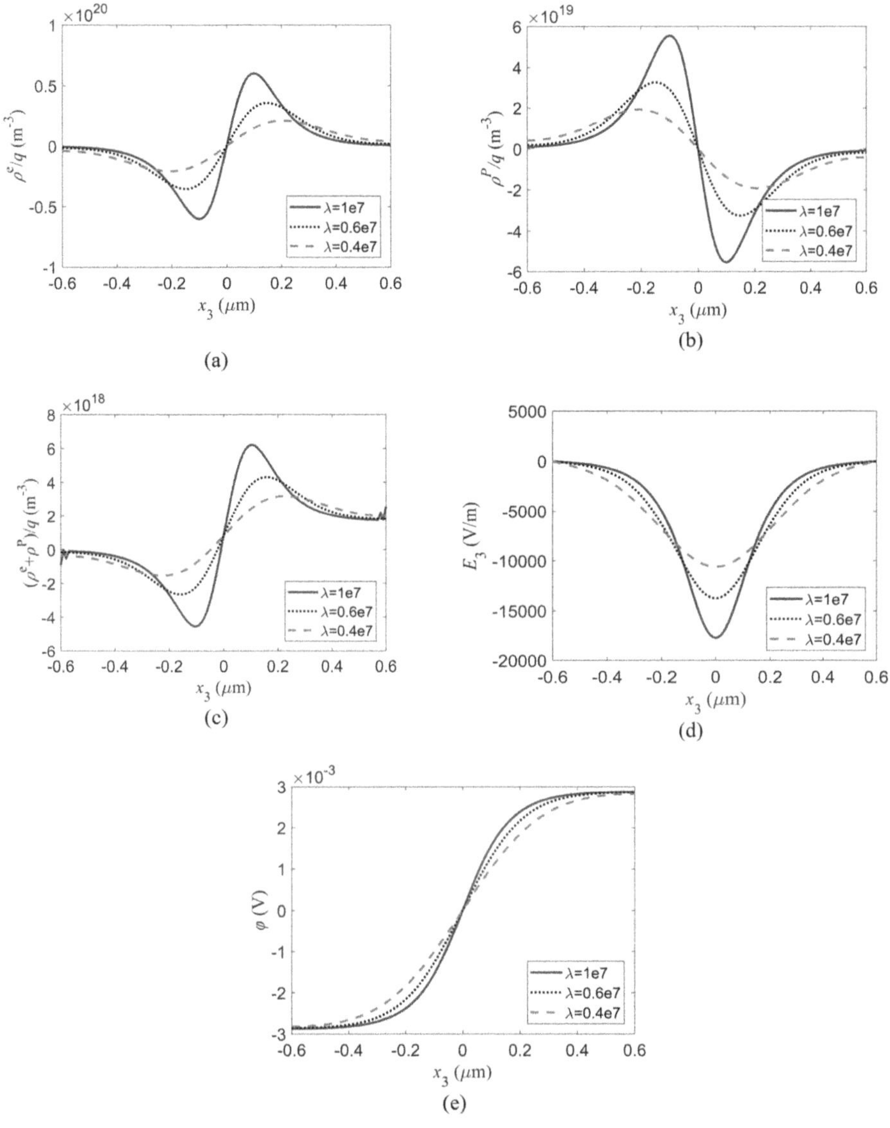

Fig. 3.5 Effects of λ. (**a**) Doping and mobile charge ρ^e/q. (**b**) Polarization charge ρ^P/q. (**c**) Total charge $(\rho^e + \rho^P)/q$. (**d**) Electric field $E_3(x_3)$. (**e**) Electric potential $\varphi(x_3)$

of P_3 at $x_3 = \pm L$ which in this specific example are not considered because for a PN junction the electric field, electric displacement, and polarization vanish sufficiently far away from the junction [3]. (d) and (e) of Fig. 3.5 show clearly that a PN junction has formed with a typical built-in electric field and a built-in potential. As λ increases, all fields change more rapidly as expected.

3.5 Periodic Doping

In this section we consider a rod with a periodically varying doping profile described by [2]:

$$\begin{aligned} N_A^-(x_3) &= a_1 + b_1 \sin \lambda x_3, \\ N_D^+(x_3) &= a_2 + b_2 \sin \lambda x_3. \end{aligned} \tag{3.62}$$

Hence

$$\begin{aligned} &N_D^+(x_3) - N_A^-(x_3) = a + b \sin \lambda x_3, \\ &a = a_2 - a_1, \quad b = b_2 - b_1. \end{aligned} \tag{3.63}$$

Then, for the particular solution of Eq. (3.19), we have

$$\varphi^p = \frac{q}{\varepsilon_{33}} \left[\frac{a}{k^2} + \frac{b}{k^2 + \lambda^2} \sin \lambda x_3 \right]. \tag{3.64}$$

In this case, it can be shown [2] that p^0 and n^0, i.e., the values of p and n at zero potential, are determined by the following equations:

$$\begin{aligned} p^0 - n^0 + a &= 0, \\ p^0 &= \frac{n^0 a_1}{2n^0 - a_2}. \end{aligned} \tag{3.65}$$

In the special case when $a_2 = a_1$, Eq. (3.63) implies that $a = 0$. Then, from Eq. (3.65), we obtain $p^0 = n^0 = a_1$. For a numerical example, we consider the case when $a_1 = 10^{21}\text{m}^{-3}$, $a_2 = a_1$, $b_1 = 0.2a_1$, $b_2 = -b_1$, and $\lambda = 0.2k$. Then $2\pi/\lambda \cong 3$ μm which can be viewed as some wavelength of the doping. We choose $2L = 30$ μm which is about ten times the doping wavelength $2\pi/\lambda$. The doping and mobile charge density ρ^e whose definition is in Eq. (3.42), the electric field $E_3(x_3)$, and the axial strain $S_3(x_3)$ in this case are shown in Fig. 3.6. They are essentially periodic except at $x_3 = \pm L$ where there are some edge effects. ρ^e has some concentration near the ends.

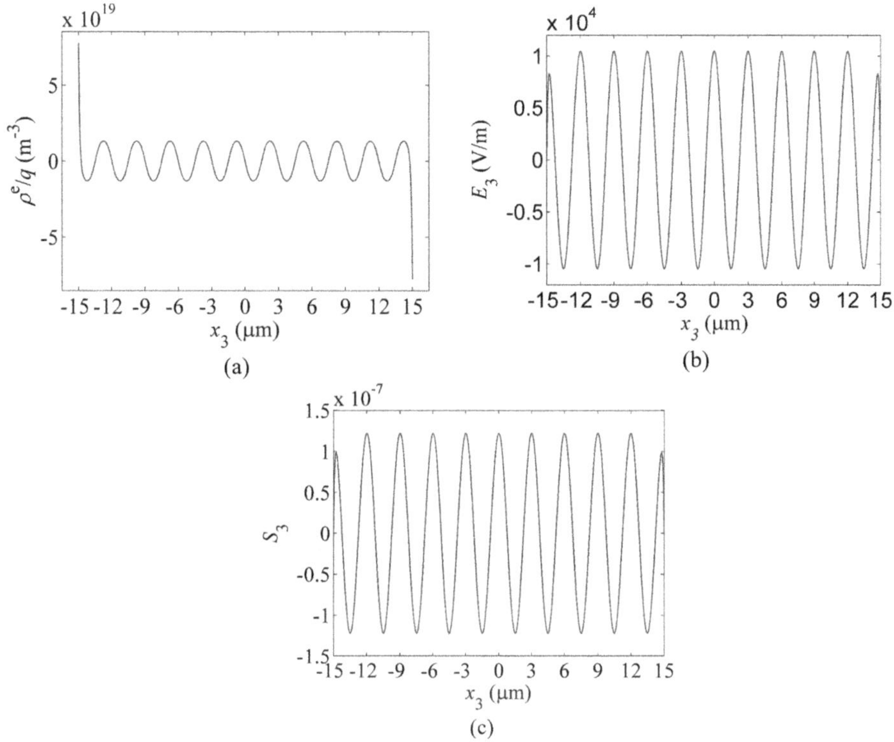

Fig. 3.6 Fields produced by periodic doping. (**a**) ρ^e/q. (**b**) $E_3(x_3)$. (**c**) $S_3(x_3)$

3.6 Linear Extension by End Forces

Consider the static extension of the rod shown in Fig. 3.7 [4]. It is within $|x_3|<L$. The cross-sectional area is A. The rod is under the action of a pair of equal and opposite axial forces f which produces an axial stress T. We are interested in the effects of T on the carrier distribution along the rod.

From Eqs. $(3.5)_1$ and (1.10), we have

$$\begin{aligned} &\frac{\partial T_3}{\partial x_3} = 0, \quad \frac{\partial D_3}{\partial x_3} = q(\Delta p - \Delta n), \\ &\frac{\partial J_3^p}{\partial x_3} = 0, \quad \frac{\partial J_3^n}{\partial x_3} = 0, \end{aligned} \tag{3.66}$$

where, following Eqs. (1.7) and (1.8),

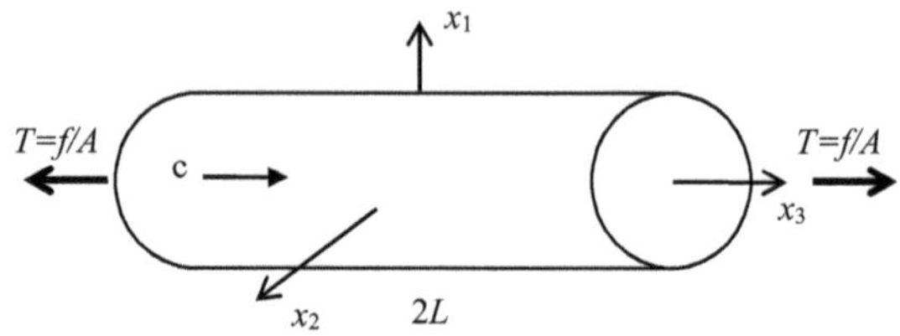

Fig. 3.7 A piezoelectric semiconductor rod of crystals of class (6mm)

$$\begin{aligned} \Delta p &= p - p_0, \quad \Delta n = n - n_0, \\ p_0 &= N_A^-, \quad n_0 = N_D^+. \end{aligned} \tag{3.67}$$

We consider the case of uniform doping. p_0 and n_0 are constants. The linearized constitutive relations are from Eqs. (3.2) and (1.9):

$$\begin{aligned} T_3 &= \bar{c}_{33} S_3 - \bar{e}_{33} E_3, \\ D_3 &= \bar{e}_{33} S_3 + \bar{\varepsilon}_{33} E_3, \end{aligned} \tag{3.68}$$

$$\begin{aligned} J_3^p &= q p_0 \mu_{33}^p E_3 - q D_{33}^p \frac{\partial (\Delta p)}{\partial x_3}, \\ J_3^n &= q n_0 \mu_{33}^n E_3 + q D_{33}^n \frac{\partial (\Delta n)}{\partial x_3}, \end{aligned} \tag{3.69}$$

where, from Eq. (3.6),

$$S_3 = u_{3,3}, \quad E_3 = -\varphi_3. \tag{3.70}$$

Consider an electrically isolated rod. There are no concentrated free charges at its ends, and there are no currents flowing in or out of the rod at its ends. In this case the boundary conditions are

$$T_3(\pm L) = T, \ D_3(\pm L) = 0, \tag{3.71}$$

$$J_3^p(\pm L) = 0, \ J_3^n(\pm L) = 0. \tag{3.72}$$

The rod is assumed to be electrically neutral at the reference state. Therefore Δn and Δp must satisfy the following charge neutrality conditions:

$$\int_{-L}^{L} \Delta n dx = 0, \ \int_{-L}^{L} \Delta p dx = 0. \tag{3.73}$$

Equation (3.73) is equivalent to one condition because of the charge equation in Eq. (3.66) and the charge boundary conditions in Eq. (3.71). To determine the mechanical displacement and the electric potential uniquely, we set

$$u_3(0) = 0, \quad \varphi(0) = 0. \tag{3.74}$$

The solution can be obtained in a systematic and straightforward manner.

The electromechanical fields in the rod are found to be

$$\begin{aligned}
\varphi &= \frac{\bar{e}_{33}T}{\varepsilon_{33}^T \bar{c}_{33} k \cosh kL} \sinh kx_3, \\
E_3 &= -\frac{\bar{e}_{33}T}{\varepsilon_{33}^T \bar{c}_{33} \cosh kL} \sinh kx_3, \\
D_3 &= -\frac{\bar{e}_{33}T}{\bar{c}_{33} \cosh kL} \cosh kx_3 + \bar{e}_{33}\frac{T}{\bar{c}_{33}},
\end{aligned} \tag{3.75}$$

$$\begin{aligned}
\Delta n &= n_0 \frac{\mu_{33}^n}{D_{33}^n} \frac{\bar{e}_{33}T}{\varepsilon_{33}^T \bar{c}_{33} k \cosh kL} \sinh kx_3, \\
\Delta p &= -p_0 \frac{\mu_{33}^p}{D_{33}^p} \frac{\bar{e}_{33}T}{\varepsilon_{33}^T \bar{c}_{33} k \cosh kL} \sinh kx_3,
\end{aligned} \tag{3.76}$$

$$\begin{aligned}
u_3 &= -\frac{\bar{e}_{33}^2 T}{\varepsilon_{33}^T \bar{c}_{33}^2 k \cosh kL} \sinh kx_3 + \frac{T}{\bar{c}_{33}} x_3, \\
S_3 &= -\frac{\bar{e}_{33}^2 T}{\varepsilon_{33}^T \bar{c}_{33}^2 \cosh kL} \cosh kx_3 + \frac{T}{\bar{c}_{33}}, \\
T_3 &= T,
\end{aligned} \tag{3.77}$$

where

$$k^2 = \frac{q}{\varepsilon_{33}^T}\left(\frac{\mu_{33}^p}{D_{33}^p} p_0 + \frac{\mu_{33}^n}{D_{33}^n} n_0\right) = \frac{q^2(p_0 + n_0)}{\varepsilon_{33}^T k_B T}. \tag{3.78}$$

The fields are either symmetric or antisymmetric about the center of the rod. The development of the mechanical displacement, stress, strain, electric potential, electric field, and electric displacement is a consequence of elasticity and piezoelectricity. When the rod is a piezoelectric dielectric, these fields are simple linear functions or constants. Now they are affected by the mobile charges and are described by hyperbolic functions. It can be verified that when $p_0 \to 0$, $n_0 \to 0$, and hence $k \to 0$, the above fields reduce to the fields for a piezoelectric dielectric rod under an axial force.

For numerical results consider an n-type ($p \cong 0$) ZnO rod with $2L = 1.2$ μm, $n_0 = N_D^+ = 10^{21}\text{m}^{-3}$, and $f = 8.5$ nN or lower which produces and axial stress $T = 3.272 \times 10^5$ N/m^2 or lower. For these data, Fig. 3.8 shows the most basic behavior of the rod as a piezoelectric semiconductor, i.e., under the action of an axial force, the electrons in the rod redistribute themselves and assume a certain distribution. Numerical calculation also shows that when n_0 increases, k increases, the hyperbolic functions have larger values, and they change more rapidly near the ends.

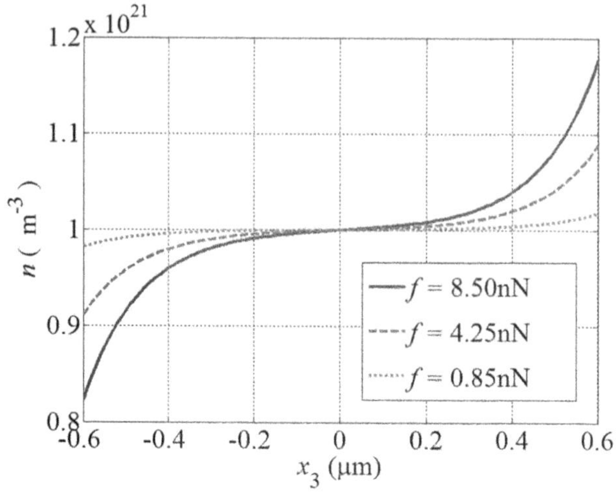

Fig. 3.8 Axial distribution of electron concentration

Δp and Δn are antisymmetric about $x_3 = 0$. We calculate the change of mobile charges in half of the rod denoted by Q^e as

$$\begin{aligned} Q^e &= \int_{-L}^{0} \rho^e A dx_3 = \int_{-L}^{0} q(\Delta p - \Delta n) A dx_3 \\ &= AT \frac{\bar{e}_{33}}{\bar{c}_{33}} \left(1 - \frac{1}{\cosh kL}\right), \\ \frac{Q^e}{AT} &= \frac{\bar{e}_{33}}{\bar{c}_{33}} \left(1 - \frac{1}{\cosh kL}\right). \end{aligned} \tag{3.79}$$

$Q^e/(AT)$ can be used as a measure of the strength of the coupling of interest, i.e., the effect of f on the redistribution of mobile charges in the rod. When p_0 or n_0 increases, k increases, coshkL increases, and Q^e increases as expected.

3.7 Electrically Nonlinear Extension by End Forces

Consider the piezoelectric semiconductor rod shown in Fig. 3.9. It is under an axial force F. F produces an axial stress $f = F/A$. The difference from the previous section is that the electrical nonlinearity in the drift current is considered in this section [5]. f in this section has a different meaning from that in the previous section.

To simplify the notation, we denote

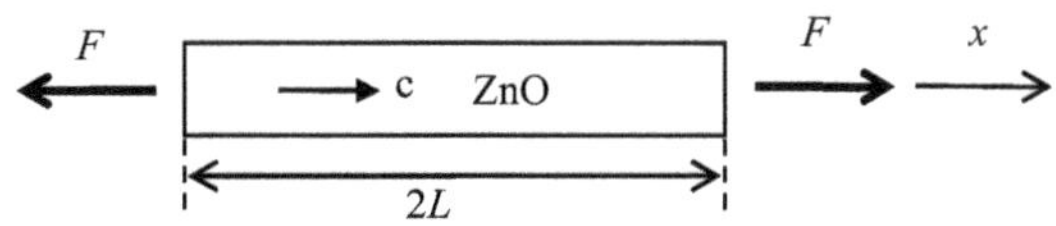

Fig. 3.9 A ZnO rod in extension

$$\begin{aligned} &x_3 = x, \quad u_3 = u, \quad S_{33} = S, \quad T_{33} = T, \\ &E_3 = E, \quad D_3 = D, \quad J_3^p = J^p, \quad J_3^n = J^n, \\ &\bar{c}_{33} = c, \quad \bar{e}_{33} = e, \quad \bar{\varepsilon}_{33} = \varepsilon, \\ &\mu_{33}^p = \mu^p, \quad D_{33}^p = D^p, \quad \mu_{33}^n = \mu^n, \quad D_{33}^n = D^n. \end{aligned} \tag{3.80}$$

From Eqs. (3.5), (3.2), (3.4), and (3.6)

$$\begin{aligned} &\frac{\partial T}{\partial x} = \rho \ddot{u}, \\ &\frac{\partial D}{\partial x} = q\left(p - n + N_D^+ - N_A^-\right) , \end{aligned} \tag{3.81}$$

$$\begin{aligned} &\frac{\partial J^p}{\partial x} = -q\frac{\partial p}{\partial t}, \\ &\frac{\partial J^n}{\partial x} = q\frac{\partial n}{\partial t}, \end{aligned} \tag{3.82}$$

$$\begin{aligned} &T = cS - eE, \\ &D = eS + \varepsilon E, \end{aligned} \tag{3.83}$$

$$\begin{aligned} &J^p = qp\mu^p E - qD^p\frac{\partial p}{\partial x}, \\ &J^n = qn\mu^n E + qD^n\frac{\partial n}{\partial x}, \end{aligned} \tag{3.84}$$

$$S = \frac{\partial u}{\partial x}, \quad E = -\frac{\partial \varphi}{\partial x}. \tag{3.85}$$

Consider the case of an n-doped rod with $p \cong 0$ and $N_A^- \cong 0$. For static extension, all fields are functions of x only. From Eqs. (3.83) and (3.85) we obtain

$$\begin{aligned} &T = c\frac{du}{dx} + e\frac{d\varphi}{dx}, \\ &D = e\frac{du}{dx} - \varepsilon\frac{d\varphi}{dx}. \end{aligned} \tag{3.86}$$

The substitution of Eq. (3.86) into the static form of Eq. (3.81) gives

$$\begin{aligned} c\frac{d^2u}{dx^2} + e\frac{d^2\varphi}{dx^2} &= 0, \\ e\frac{d^2u}{dx^2} - \varepsilon\frac{d^2\varphi}{dx^2} &= q\left(-n + N_D^+\right). \end{aligned} \tag{3.87}$$

Equation (3.87) can be rewritten as

$$\begin{aligned} \frac{d^2u}{dx^2} &= -\frac{e}{c}\frac{d^2\varphi}{dx^2}, \\ n &= \frac{\varepsilon_{33}^T}{q}\frac{d^2\varphi}{dx^2} + N_D^+. \end{aligned} \tag{3.88}$$

In the following we restrict ourselves to the case of uniform doping with a constant N_D^+. From the static form of Eq. (3.82), J^n is a constant. We consider an electrically isolated rod with $J^n = 0$ at its ends. Hence J^n is identically zero throughout the rod. Then, from Eqs. $(3.84)_2$ and (3.85), we obtain

$$-n\mu^n\frac{d\varphi}{dx} + D^n\frac{dn}{dx} = 0. \tag{3.89}$$

In summary, we have three equations in Eqs. (3.88) and (3.89) for φ, n, and u. The boundary conditions are

$$T(\pm L) = f, \quad D(\pm L) = 0, \quad J^n(\pm L) = 0. \tag{3.90}$$

The charge neutrality condition is

$$\int_{-L}^{L}\left(-n + N_D^+\right)\, dx = 0, \tag{3.91}$$

which in this case of electrons only without holes is not an independent condition and is implied by Eq. $(3.81)_2$ and the boundary conditions on D in Eq. (3.90). It is kept here formally only. To make the mechanical displacement and the electric potential fields unique, we need to impose the following conditions:

$$u(0) = 0, \quad \varphi(0) = 0. \tag{3.92}$$

We look for a perturbation solution in the following form in terms of zero-, first-, and second-order fields:

$$
\begin{aligned}
n &\cong n^{(0)} + n^{(1)} + n^{(2)}, \\
\varphi &\cong \varphi^{(1)} + \varphi^{(2)}, \\
u &\cong u^{(1)} + u^{(2)}.
\end{aligned} \tag{3.93}
$$

Substituting Eq. (3.93) into the above equations, we obtain a series of perturbation problems. The reference state with $n^{(0)} = N_D^+$ is the solution to the zero-order problem in which all other fields vanish. The first-order problem is

$$
\begin{aligned}
n^{(1)} &= \frac{\varepsilon_{33}^T}{q} \frac{d^2\varphi^{(1)}}{dx^2}, \\
\frac{d^2u^{(1)}}{dx^2} &= -\frac{e}{c} \frac{d^2\varphi^{(1)}}{dx^2}, \\
-n^{(0)}\mu^n \frac{d\varphi^{(1)}}{dx} &+ D^n \frac{dn^{(1)}}{dx} = 0,
\end{aligned} \tag{3.94}
$$

and

$$
\begin{aligned}
&T^{(1)}(\pm L) = f, \quad D^{(1)}(\pm L) = 0, \\
&\int_{-L}^{L} n^{(1)} \, dx = 0, \\
&u^{(1)}(0) = 0, \quad \varphi^{(1)}(0) = 0.
\end{aligned} \tag{3.95}
$$

The solution to the first-order problem was obtained in the previous section. The electric potential, electron concentration, and mechanical displacement fields are

$$
\varphi^{(1)} = \frac{ef}{\varepsilon_{33}^T ck \cosh kL} \sinh kx, \tag{3.96}
$$

$$
n^{(1)} = N_D^+ \frac{\mu^n}{D^n} \frac{ef}{\varepsilon_{33}^T ck \cosh kL} \sinh kx, \tag{3.97}
$$

$$
u^{(1)} = -\frac{e^2 f}{\varepsilon_{33}^T c^2 k \cosh kL} \sinh kx + \frac{f}{c} x, \tag{3.98}
$$

where

$$
k^2 = \frac{q}{\varepsilon_{33}^T} \frac{\mu^n}{D^n} N_D^+. \tag{3.99}
$$

Clearly, the first-order perturbation solution represents a linear solution proportional to f. The second-order problem is given by

$$\begin{aligned} n^{(2)} &= \frac{\varepsilon_{33}^T}{q} \frac{d^2 \varphi^{(2)}}{dx^2}, \\ \frac{d^2 u^{(2)}}{dx^2} &= -\frac{e}{c} \frac{d^2 \varphi^{(2)}}{dx^2}, \\ -n^{(0)} \mu^n \frac{d\varphi^{(2)}}{dx} + D^n \frac{dn^{(2)}}{dx} &= n^{(1)} \mu^n \frac{d\varphi^{(1)}}{dx}, \end{aligned} \tag{3.100}$$

and

$$\begin{aligned} & T^{(2)}(\pm L) = 0, \quad D^{(2)}(\pm L) = 0, \\ & \int_{-L}^{L} n^{(2)} \, dx = 0, \\ & u^{(2)}(0) = 0, \quad \varphi^{(2)}(0) = 0. \end{aligned} \tag{3.101}$$

Through some algebra the second-order solution is found to be

$$\begin{aligned} \varphi^{(2)} = &- \frac{qm \cosh kL}{6k^2 \varepsilon_{33}^T \sinh kL} [\sinh k(x+L) - \sinh k(x-L)] \\ &+ \frac{qm}{12k^2 \varepsilon_{33}^T} \cosh 2kx - \frac{q}{k^2 \varepsilon_{33}^T} \left[\frac{m}{12} - \frac{m \cosh kL}{3} \right], \end{aligned} \tag{3.102}$$

$$\begin{aligned} n^{(2)} = &- \frac{m \cosh kL}{6 \sinh kL} [\sinh k(x+L) - \sinh k(x-L)] \\ &+ \frac{m}{3} \cosh 2kx, \end{aligned} \tag{3.103}$$

$$\begin{aligned} u^{(2)} = &\frac{e}{c} \frac{qm \cosh kL}{6k^2 \varepsilon_{33}^T \sinh kL} [\sinh k(x+L) - \sinh k(x-L)] \\ &- \frac{e}{c} \frac{qm}{12k^2 \varepsilon_{33}^T} \cosh 2kx + \frac{e}{c} \frac{q}{k^2 \varepsilon_{33}^T} \left[\frac{m}{12} - \frac{m \cosh kL}{3} \right], \end{aligned} \tag{3.104}$$

where

$$m = N_D^+ \left(\frac{\mu^n}{D^n} \right)^2 \left(\frac{ef}{\varepsilon_{33}^T ck \cosh kL} \right)^2. \tag{3.105}$$

It can be seen that the second-order solution is quadratic in f through m.

As a numerical example, consider a ZnO rod with $2L = 1.2$ μm and $n_0 = N_D^+ = 10^{21} \mathrm{m}^{-3}$. Figure 3.10 shows the comparison of the first-order linear solution described by $n \cong n^{(0)} + n^{(1)}$ with the second-order nonlinear solution when the applied axial stress $f = 2 \times 10^6$ N/m^2. When f is small, the first-order solution is expected to be accurate. Our numerical tests indeed show that when $f < 0.6 \times 10^6$ N/m^2, the first- and second-order solutions become indistinguishable when plotted in the scale of Fig. 3.10. When the axial stress is increased to $f = 2 \times 10^6$ N/m^2, the

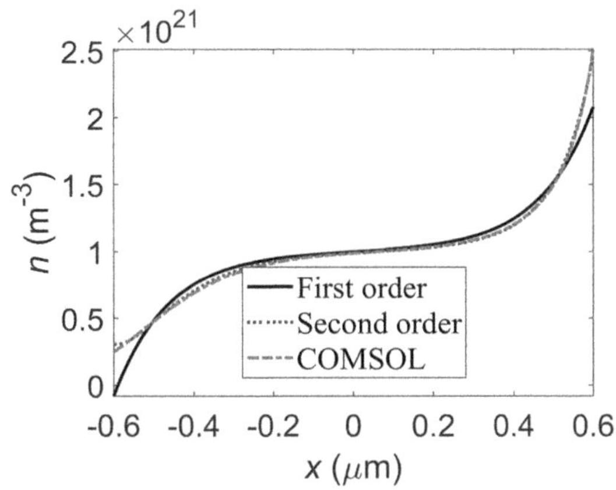

Fig. 3.10 Comparison of first-order (linear), second-order (nonlinear), and COMSOL (nonlinear) solutions of electron concentration distribution

first- and second-order solutions shown in Fig. 3.10 become significantly different. The first-order solution represented by the solid line is antisymmetric about the middle of the rod. However, the second-order solution loses the antisymmetry. This is a fundamental and qualitative difference. In fact, the electromechanical fields described by the second-order solution lose the symmetry or antisymmetry of the first-order solution in general [5]. A finite element numerical solution using the commercial software COMSOL is also provided in the figure which is very close to the second-order perturbation solution.

From Eq. (3.21), the polarization and the effective polarization charge density can be calculated from

$$\begin{aligned} P_3 &= D_3 - \varepsilon_0 E_3, \\ \rho^P &= -P_{k,k} = -P_{3,3}. \end{aligned} \tag{3.106}$$

They are found to be

$$\begin{aligned} P^{(1)} =& D^{(1)} - \varepsilon_0 E^{(1)}, \\ \rho^{P(1)} =& -\nabla \cdot P^{(1)} \\ =& \frac{kef}{c\cosh kL}\sinh kx - \varepsilon_0 \frac{kef}{\varepsilon_{33}^T c\cosh kL}\sinh kx, \end{aligned} \tag{3.107}$$

$$\begin{aligned} \rho^{P(2)} =& -\nabla \cdot P^{(2)} \\ =& -\frac{qm\cosh kL}{6\sinh kL}[\sinh k(x+L) - \sinh k(x-L)] \\ &+ \frac{\varepsilon_0 qm\cosh kL}{6\varepsilon_{33}^T\sinh kL}[\sinh k(x+L) - \sinh k(x-L)] \\ &+ \frac{qm}{3}\cosh 2kx - \frac{\varepsilon_0 qm}{3\varepsilon_{33}^T}\cosh 2kx. \end{aligned} \tag{3.108}$$

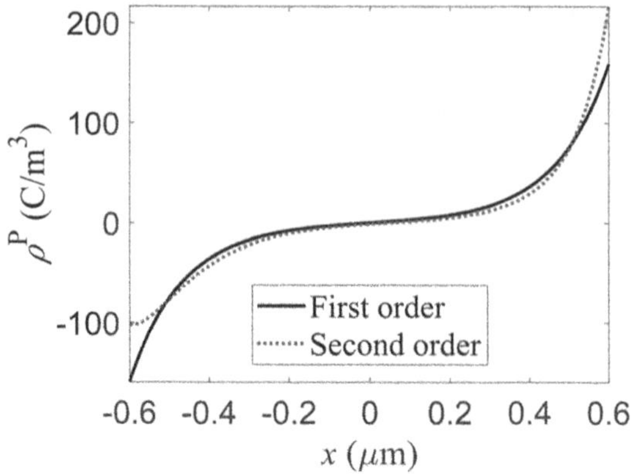

Fig. 3.11 Comparison of first-order (linear) and second-order (nonlinear) solutions of effective polarization charge density

ρ^P is plotted in Fig. 3.11. It is induced by the extensional force through piezoelectric coupling and is screened by the mobile charges in a semiconductor. There also exist concentrated effective polarization charges at the two ends of the rod.

3.8 Local Extension

Consider the piezoelectric semiconductor rod of crystals of class (6mm) shown in Fig. 3.12. Its c-axis is along $x = x_3$. The rod is unbounded along x and is under a local pair of concentrated forces [6]. We consider linear extension.

We follow the notation of Eq. (3.80). p_0 and n_0 are defined by Eq. (1.8). The governing equations are from Eqs. $(1.1)_1$ with $i = 3$, (1.10), (3.2), (1.9) with $i = 3$, and (3.6):

$$\begin{aligned} &\frac{\partial T}{\partial x} + f = \rho \ddot{u}, \\ &\frac{\partial D}{\partial x} = q(\Delta p - \Delta n) \ , \end{aligned} \tag{3.109}$$

$$\begin{aligned} &q\frac{\partial(\Delta p)}{\partial t} = -\frac{\partial J^p}{\partial x}, \\ &q\frac{\partial(\Delta n)}{\partial t} = \frac{\partial J^n}{\partial x}, \end{aligned} \tag{3.110}$$

$$\begin{aligned} T &= cS - eE, \\ D &= eS + \varepsilon E, \end{aligned} \tag{3.111}$$

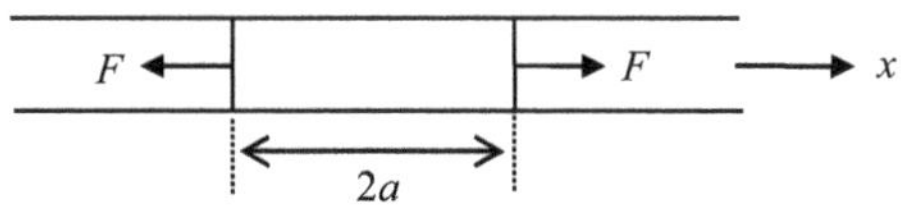

Fig. 3.12 A piezoelectric semiconductor rod under a local pair of concentrated forces

$$
\begin{aligned}
J^p &= qp_0\mu^p E - qD^p \frac{\partial(p_0 + \Delta p)}{\partial x}, \\
J^n &= qn_0\mu^n E + qD^n \frac{\partial(n_0 + \Delta n)}{\partial x},
\end{aligned}
\tag{3.112}
$$

$$
S = \frac{\partial u}{\partial x}, \quad E = -\frac{\partial \varphi}{\partial x}, \tag{3.113}
$$

where $f = f_3$. The definitions of Δp and Δn are given in Eq. (1.7). We consider uniform p_0 and n_0. With successive substitutions, Eqs. (3.109) and (3.110) can be written as

$$
\begin{aligned}
& c\frac{\partial^2 u}{\partial x^2} + e\frac{\partial^2 \varphi}{\partial x^2} + f = \rho\ddot{u}, \\
& e\frac{\partial^2 u}{\partial x^2} - \varepsilon\frac{\partial^2 \varphi}{\partial x^2} = q(\Delta p - \Delta n),
\end{aligned}
\tag{3.114}
$$

$$
\begin{aligned}
\frac{\partial}{\partial t}(\Delta p) &= p_0\mu^p \frac{\partial^2 \varphi}{\partial x^2} + D^p \frac{\partial^2(\Delta p)}{\partial x^2}, \\
\frac{\partial}{\partial t}(\Delta n) &= -n_0\mu^n \frac{\partial^2 \varphi}{\partial x^2} + D^n \frac{\partial^2(\Delta n)}{\partial x^2}.
\end{aligned}
\tag{3.115}
$$

For static problems, Eqs. (3.114) and (3.115) reduce to

$$
\begin{aligned}
& c\frac{d^2 u}{dx^2} + e\frac{d^2 \varphi}{dx^2} + f = 0, \\
& e\frac{d^2 u}{dx^2} - \varepsilon\frac{d^2 \varphi}{dx^2} = q(\Delta p - \Delta n),
\end{aligned}
\tag{3.116}
$$

$$
\begin{aligned}
0 &= p_0\mu^p \frac{d^2 \varphi}{dx^2} + D^p \frac{d^2(\Delta p)}{dx^2}, \\
0 &= -n_0\mu^n \frac{d^2 \varphi}{dx^2} + D^n \frac{d^2(\Delta n)}{dx^2}.
\end{aligned}
\tag{3.117}
$$

Equations (3.116) and (3.117) can be partially decoupled into

$$\frac{d^2(\Delta p-\Delta n)}{dx^2}=k^2(\Delta p-\Delta n)+\frac{ek^2}{cq}f, \tag{3.118}$$

$$\frac{d^2\varphi}{dx^2}=-\frac{q}{k^2\hat{\varepsilon}}\frac{d^2(\Delta p-\Delta n)}{dx^2}, \tag{3.119}$$

$$\frac{d^2u}{dx^2}=-\frac{e}{c}\frac{d^2\varphi}{dx^2}-\frac{f}{c}, \tag{3.120}$$

$$\frac{d^2(\Delta p)}{dx^2}=-\frac{p_0\mu^p}{D^p}\frac{d^2\varphi}{dx^2}, \tag{3.121}$$

$$\frac{d^2(\Delta n)}{dx^2}=\frac{n_0\mu^n}{D^n}\frac{d^2\varphi}{dx^2}, \tag{3.122}$$

where only one of Eqs. (3.121) and (3.122) is necessary, and

$$\begin{aligned} k^2&=\left(\frac{p_0\mu^p}{D^p}+\frac{n_0\mu^n}{D^n}\right)\frac{q}{\hat{\varepsilon}},\\ \hat{\varepsilon}&=\varepsilon+\frac{e^2}{c}=\varepsilon_{33}^T. \end{aligned} \tag{3.123}$$

Consider a ZnO rod doped into an n-type semiconductor with $p\cong 0$. We assume there is no body force so that $f=0$. We need to find solutions for each of the three regions in $x<-a$, $|x|<a$, and $x>a$, respectively, and apply boundary conditions and continuity (or jump) conditions. The general solution to Eqs. (3.118), (3.119), (3.120), (3.121), and (3.122) is

$$\begin{aligned} \Delta n&=C_1e^{kx}+C_2e^{-kx},\\ \varphi&=\frac{k_BT}{qn_0}\left(C_1e^{kx}+C_2e^{-kx}\right)+C_3x+C_4,\\ u&=-\frac{qe}{c\varepsilon+e^2}\frac{1}{k^2}\left[C_1e^{kx}+C_2e^{-kx}\right]+C_5x+C_6, \end{aligned} \tag{3.124}$$

where C_1–C_6 are undetermined constants. Equation (3.124) is valid for each of the three regions in Fig. 3.12 with corresponding undetermined constants for each region. All together there are 18 undetermined constants. The rod is mechanically free and electrically open at infinity. Denoting $-a^-=(-a)^-$ and $-a^+=(-a)^+$, we write the boundary and continuity conditions as

$$T(-\infty)=0,\quad D(-\infty)=0,\quad J^n(-\infty)=0, \tag{3.125}$$

$$\begin{aligned} &u(-a^-) = u(-a^+), \quad T(-a^+) - T(-a^-) = F/A, \\ &\varphi(-a^-) = \varphi(-a^+), \quad D(-a^-) = D(-a^+), \\ &n(-a^-) = n(-a^+), \quad J^n(-a^-) = J^n(-a^+), \end{aligned} \tag{3.126}$$

$$\begin{aligned} &u(a^-) = u(a^+), \quad T(a^+) - T(a^-) = -F/A, \\ &\varphi(a^-) = \varphi(a^+), \quad D(a^-) = D(a^+), \\ &n(a^-) = n(a^+), \quad J^n(a^-) = J^n(a^+), \end{aligned} \tag{3.127}$$

$$T(+\infty) = 0, \quad D(+\infty) = 0, \quad J^n(+\infty) = 0. \tag{3.128}$$

Equations (3.125), (3.126), (3.127), and (3.128) are 18 conditions, but some of them are not independent. For example, $J^n(-\infty) = 0$, J^n is continuous at $\pm a$, and $\partial J^n/\partial x = 0$ imply that $J^n(\infty) = 0$. Hence, to determine the displacement and potential fields uniquely, we need to choose $-\infty$ as a reference and set

$$u(-\infty) = 0, \quad \varphi(-\infty) = 0. \tag{3.129}$$

The substitution of Eq. (3.124) for different regions into Eqs. (3.125), (3.126), (3.127), (3.128), and (3.129) results in a system of linear equations for the undetermined constants. They are solved on a computer using MATLAB. Then various electromechanical fields can be calculated.

(a)–(c) of Fig. 3.13 show the effective polarization charge density as defined by Eq. (3.21), the electron concentration perturbation, and the electric potential produced by two different values of F when $n_0 = 10^{21}\text{m}^{-3}$ and $a = 600$ nm. There also exist concentrated polarization charges at $x = \pm a$. For the purpose of mechanically manipulating the electrical behavior of the rod, the most basic effect of F is that it produces a potential well followed by a potential barrier as shown in (c). With the presence of the potential well and barrier, a mobile charge cannot travel through the rod unless it has sufficient velocity or kinetic energy. Therefore, a local pair of concentrated forces forbids the passing of low-energy mobile charges through the rod. We note the resemblance of the central part of (b) to Fig. 3.8 for the electron distribution in the extension of a rod by end forces. (d) shows the effect of F on the current-voltage relation from a nonlinear numerical analysis by COMSOL [7]. It can be seen that when the applied voltage V is low, no current can flow through the rod in either direction. When a positive V exceeds a stress-dependent critical value, currents begin to flow in one direction of the rod but not in the other direction if the sign of the voltage is changed. When V exceeds a second stress-dependent critical value, currents can flow in both directions when the sign of V is changed. Therefore the local stress acts like a switch that has different behaviors in opposite directions. This provides a basic means for mechanically manipulating currents in a piezoelectric semiconductor rod.

If the lateral surface of the rod is not traction-free, then the lateral stress T_{11} and T_{22} can affect the axial distribution of the electric fields in a similar way through the piezoelectric constant e_{31}.

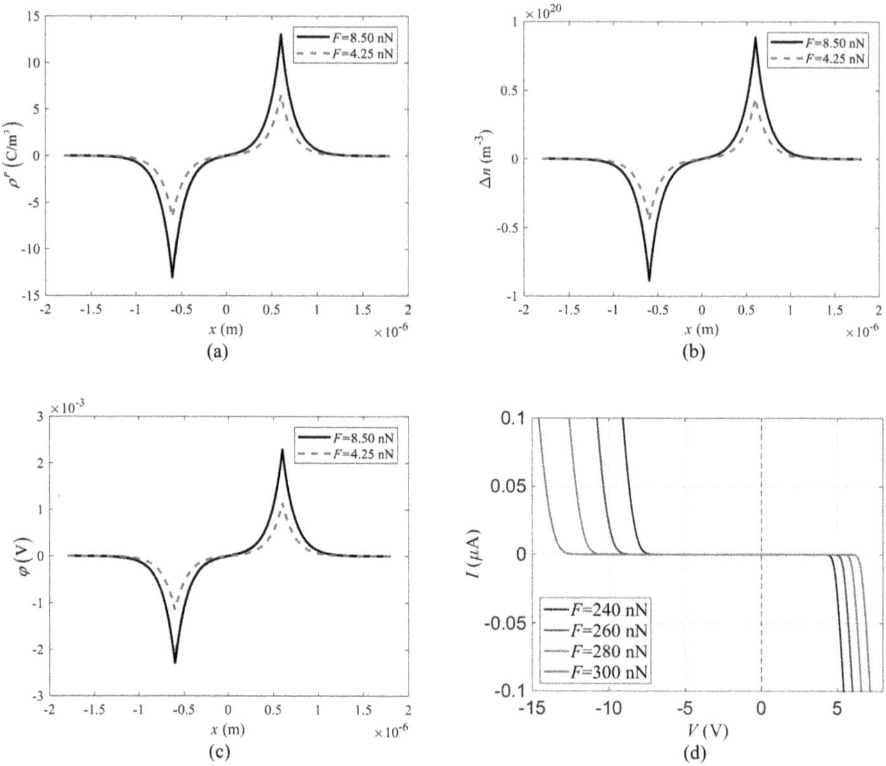

Fig. 3.13 Fields for different F. $n_0 = 10^{21}\text{m}^{-3}$. (**a**) Effective polarization charge density ρ^P. (**b**) Electron concentration perturbation Δn. (**c**) Electric potential φ. (**d**) $I-V$ curve. $V = \varphi(L) - \varphi(-L)$. $I = J^n A$

3.9 Local Extension and Compression

The fields and charge distributions obtained in the previous section by one pair of local forces can be used as building blocks for constructing more complicated electric fields and charge distributions mechanically. As an example, consider an unbounded rod under two local pairs of forces, one compressive and the other tensile as shown in Fig. 3.14 [6].

The solution to this problem is similar to that in the previous section, with more algebra because of more regions with different stresses, more undetermined constants, and more boundary and continuity conditions. The solution is not presented here, and we show the results directly in Fig. 3.15 for two different values of n_0. Our main interest is in the big potential well with two small potential barriers in (c).

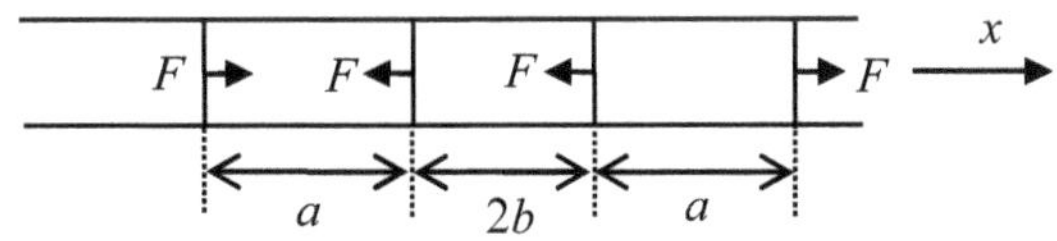

Fig. 3.14 A piezoelectric semiconductor rod under two local pairs of concentrated forces

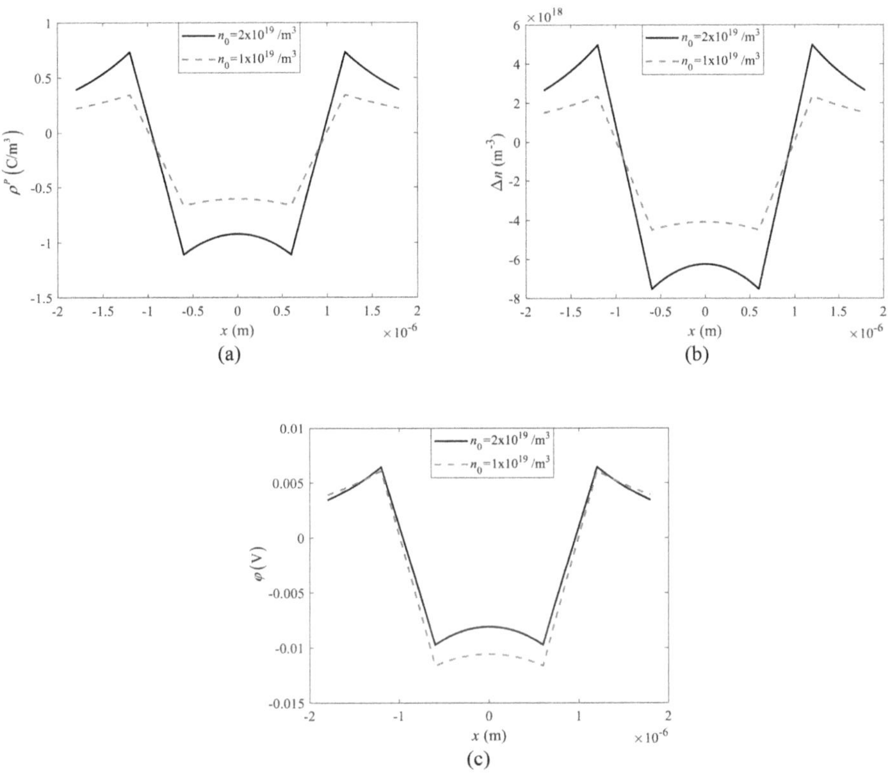

Fig. 3.15 Fields for different n_0. $F = 8.5$ nN. (**a**) Effective polarization charge density ρ^P. (**b**) Electron concentration perturbation Δn. (c) Electric potential φ

3.10 Periodic Extension and Compression

In this section we consider the effect of the distributed body force f in Eq. $(3.109)_1$ [6]. As an example, consider the case of a rod under a sinusoidal distribution of f described by $f = B \sin \xi x$ where B and ξ are constants. For an n-type rod with $p \cong 0$, it is straightforward to find the corresponding fields from Eqs. (3.118), (3.119), (3.20), (3.21), and (3.122) as

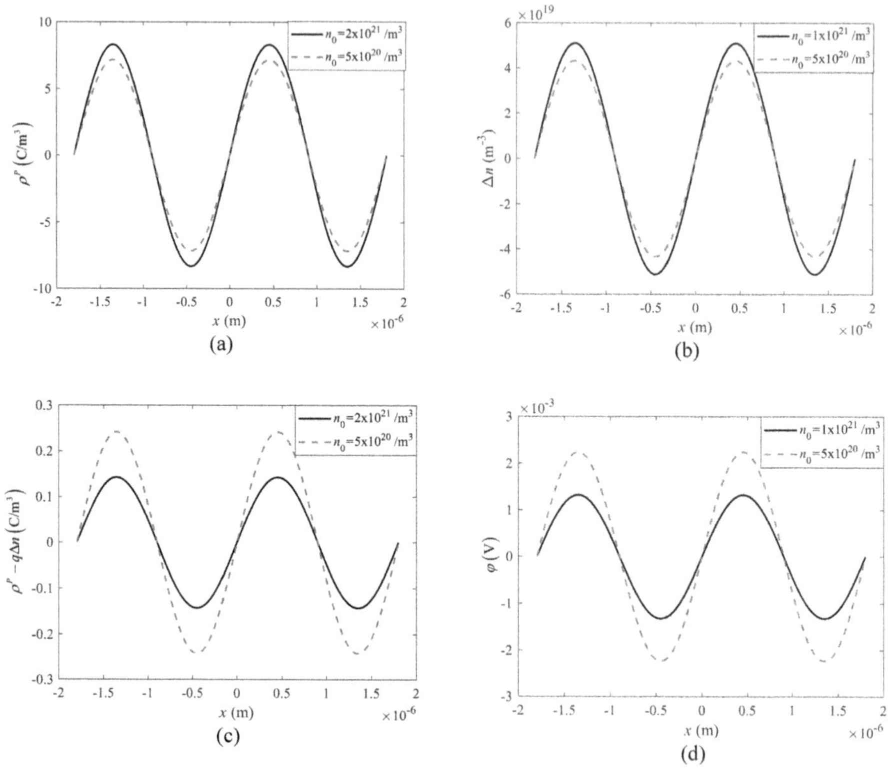

Fig. 3.16 Fields for different n_0. (**a**) Effective polarization charge density ρ^P. (**b**) Electron concentration perturbation Δn. (**c**) Screened polarization charge $\rho^P - q\Delta n$. (**d**) Electric potential φ

$$
\begin{aligned}
\Delta n &= \frac{ek^2}{cq\left(k^2+\xi^2\right)} B\sin\xi x, \\
\varphi &= \frac{k_B T}{q^2 n_0} \frac{ek^2}{c\left(k^2+\xi^2\right)} B\sin\xi x, \\
u &= -\frac{e^2}{c^2} \frac{k_B T}{q^2 n_0} \frac{k^2}{\left(k^2+\xi^2\right)} B\sin\xi x + \frac{B}{c}\frac{1}{\xi^2}\sin\xi x.
\end{aligned}
\tag{3.130}
$$

We plot the charges and potential determined by Eq. (3.130) in Fig. 3.16 when $B = 8.5536 \times 10^{11}$ N/m^3 and $\xi = 3.4907 \times 10^6$/m. The charge and potential distributions are also sinusoidal. A series of periodically alternating potential barriers, wells, and charged regions are produced. They are similar to the fields in Fig. (3.6) due to periodic doping but are produced mechanically here. In this case, for the effective polarization charge, there is a distributed ρ^P only without concentrated effective polarization charges at discontinuities because all fields are continuous everywhere. (a) to (c) show that ρ^P is significantly but not completely screened by electrons.

3.11 Harmonic Vibration

Consider the finite piezoelectric semiconductor rod shown in Fig. 3.17. The left end is fixed. The right end is under the action of a time-harmonic extensional force [8]. Both ends are electroded with ohmic contacts. There is no interface nonlinearity, and the interface resistance between the crystal and electrodes is negligible at the two ends. The two end electrodes are connected by an output circuit whose impedance is Z in time-harmonic motions. Such a rod can operate as an energy harvester for converting mechanical energy to electrical energy. As an energy harvester, the screening of the effective polarization charge by the mobile charge in a semiconductor is undesirable. It reduces the piezoelectrically generated output voltage by the mechanical load.

The governing equations are from Eqs. $(1.1)_1$ with $i = 3$, (1.10), (3.2), (1.9) with $i = 3$, and (3.6):

$$\begin{aligned} \frac{\partial T_3}{\partial x_3} &= \rho \ddot{u}_3, \\ \frac{\partial D_3}{\partial x_3} &= q(\Delta p - \Delta n) \ , \end{aligned} \tag{3.131}$$

$$\begin{aligned} q\frac{\partial}{\partial t}(\Delta p) &= -\frac{\partial J_3^p}{\partial x_3}, \\ q\frac{\partial}{\partial t}(\Delta n) &= \frac{\partial J_3^n}{\partial x_3}. \end{aligned} \tag{3.132}$$

$$\begin{aligned} T_3 &= \bar{c}_{33}S_3 - \bar{e}_{33}E_3, \\ D_3 &= \bar{e}_{33}S_3 + \bar{\varepsilon}_{33}E_3, \end{aligned} \tag{3.133}$$

$$\begin{aligned} J_3^p &= qp_0\mu_{33}^p E_3 - qD_{33}^p \frac{\partial(\Delta p)}{\partial x_3}, \\ J_3^n &= qn_0\mu_{33}^n E_3 + qD_{33}^n \frac{\partial(\Delta n)}{\partial x_3}, \end{aligned} \tag{3.134}$$

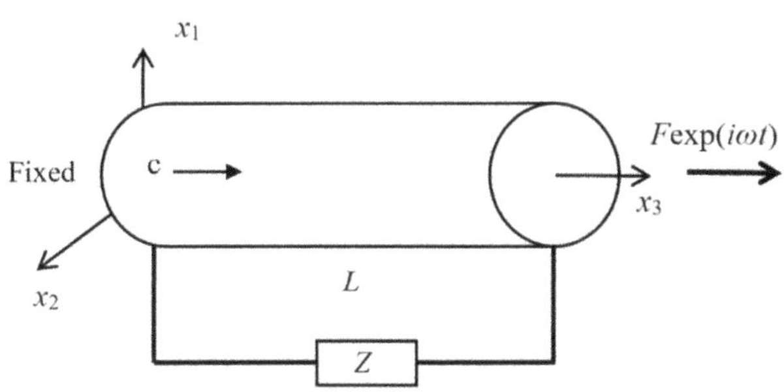

Fig. 3.17 A piezoelectric semiconductor rod under a time-harmonic force

$$S_3 = \frac{\partial u_3}{\partial x_3}, \quad E_3 = -\frac{\partial \varphi}{\partial x_3}, \tag{3.135}$$

where uniform doping has been assumed. Δp and Δn are given in Eq. (1.7). p_0 and n_0 are defined by Eq. (1.8). We consider an n-type semiconductor with $p \cong 0$. For time-harmonic vibrations described by a common time factor $\exp(i\omega t)$, with successive substitutions, we can write Eqs. (3.131) and $(3.132)_2$ as

$$\begin{aligned} &\bar{c}_{33} u_{3,33} + \bar{e}_{33} \varphi_{,33} = -\rho \omega^2 u_3, \\ &\bar{e}_{33} u_{3,33} - \bar{\varepsilon}_{33} \varphi_{,33} = q(-\Delta n), \\ &i\omega q(\Delta n) = -q n_0 \mu_{33}^n \varphi_{,33} + q D_{33}^n \frac{\partial(\Delta n)}{\partial x_3^2}. \end{aligned} \tag{3.136}$$

We look for a solution of Eq. (3.136) in the following form:

$$\begin{Bmatrix} u_3 \\ \varphi \\ \Delta n \end{Bmatrix} = \begin{Bmatrix} A \\ B \\ C \end{Bmatrix} \exp(\zeta x_3), \tag{3.137}$$

where A, B, C, and ζ are undetermined constants. Substituting Eq. (3.137) into Eq. (3.136), we obtain a system of linear homogeneous equations for A, B, and C:

$$\begin{bmatrix} \bar{c}_{33}\zeta^2 + \rho\omega^2 & \bar{e}_{33}\zeta^2 & 0 \\ \bar{e}_{33}\zeta^2 & -\bar{\varepsilon}_{33}\zeta^2 & q \\ 0 & -n_0\mu_{33}^n\zeta^2 & D_{33}^n\zeta^2 - i\omega \end{bmatrix} \begin{Bmatrix} A \\ B \\ C \end{Bmatrix} = 0. \tag{3.138}$$

For nontrivial solutions of A, B, or C, the determinant of the coefficient matrix of Eq. (3.138) has to vanish, which leads to the following cubic equation for ζ^2:

$$\begin{aligned} &-\left(\bar{c}_{33}\bar{\varepsilon}_{33} + \bar{e}_{33}^2\right) D_{33}^n \zeta^6 \\ &+ \left(i\omega \bar{c}_{33}\bar{\varepsilon}_{33} + i\omega \bar{e}_{33}^2 + \bar{c}_{33} q n_0 \mu_{33}^n - \rho\omega^2 D_{33}^n \bar{\varepsilon}_{33}\right)\zeta^4 \\ &+ \left(i\rho\omega^3 \bar{\varepsilon}_{33} + \rho\omega^2 q n_0 \mu_{33}^n\right)\zeta^2 = 0. \end{aligned} \tag{3.139}$$

Equation (3.139) has six roots for ζ. $\zeta^{(1)} = \zeta^{(2)} = 0$ is a repeated root. It is easy to show that $\zeta^{(1)} = \zeta^{(2)} = 0$ leads to

$$\begin{Bmatrix} u_3 \\ \varphi \\ \Delta n \end{Bmatrix} = B^{(1)} \begin{Bmatrix} 0 \\ x_3 \\ 0 \end{Bmatrix} + B^{(2)} \begin{Bmatrix} 0 \\ 1 \\ 0 \end{Bmatrix}, \tag{3.140}$$

where $B^{(1)}$ and $B^{(2)}$ are undetermined constants. $B^{(2)}$ represents an arbitrary constant in the potential which does not affect the electric field and is immaterial. Therefore it will be dropped in the following. Denoting the four other roots of Eq. (3.139) by $\zeta^{(3)}$ through $\zeta^{(6)}$, we write the general solution of Eq. (3.136) as

$$\begin{Bmatrix} u_3 \\ \varphi \\ \Delta n \end{Bmatrix} = B^{(1)} \begin{Bmatrix} 0 \\ x_3 \\ 0 \end{Bmatrix} + \sum_{m=3}^{6} B^{(m)} \begin{Bmatrix} \alpha^{(m)} \\ 1 \\ \beta^{(m)} \end{Bmatrix} \exp \zeta^{(m)} x_3, \tag{3.141}$$

where $B^{(3)}$ through $B^{(6)}$ are undetermined constants, and

$$\begin{aligned} \alpha^{(m)} &= -\frac{\bar{e}_{33}\left(\zeta^{(m)}\right)^2}{\bar{c}_{33}\left(\zeta^{(m)}\right)^2 + \rho\omega^2}, \\ \beta^{(m)} &= \frac{n_0 \mu_{33}^n \left(\zeta^{(m)}\right)^2}{D_{33}^n \left(\zeta^{(m)}\right)^2 - i\omega}, \quad m = 3, 4, 5, 6. \end{aligned} \tag{3.142}$$

The boundary conditions are

$$u_3(0) = 0, \quad T_3(L) = \frac{F}{S} \exp(i\omega t), \tag{3.143}$$

$$D_3(0) = 0, \quad D_3(L) = 0. \tag{3.144}$$

We also have the following circuit equation:

$$\frac{V}{Z} = J^n(L)S, \tag{3.145}$$

where $V = \varphi(L) - \varphi(0)$ is the piezoelectrically generated output voltage on Z. The substitution of Eq. (3.141) into Eqs. (3.143), (3.144), and (3.145) results in five linear equations for $B^{(1)}$ and $B^{(3)}$ through $B^{(6)}$. They are solved on a computer using MATLAB.

As a numerical example, consider a ZnO rod whose geometric parameters are that the length $L = 600$ nm, radius $a = 25$ nm, and cross-sectional area $S = \pi a^2 = 1.9625 \times 10^{-15}$ m^2. The magnitude of the end force $F = 0.5$ nN. The reference carrier concentration $n_0 = N_D^+ = 10^{22}$ m^{-3}. For the mechanically forced vibration we are considering, to introduce some mechanical damping into the structure, we use a complex elastic constant $\bar{c}_{33}\left(1 + iQ^{-1}\right)$ in our numerical calculation which can describe viscous damping in time-harmonic motions. Q is the material quality factor. For convenience we introduce

$$\omega_0^{(n)} = \frac{(2n-1)\pi}{2L}\sqrt{\frac{\bar{c}_{33}}{\rho}}, \quad n = 1, 2, 3, \cdots, \tag{3.146}$$

which are the extensional resonance frequencies of the corresponding elastic dielectric rod. We also introduce

$$Z_0 = \frac{1}{i\omega_0^{(1)} C_0}, \quad C_0 = \bar{\varepsilon}_{33}\frac{S}{L} \tag{3.147}$$

as a unit for the impedance Z of the output circuit.

We calculate the open-circuit vibration characteristics when $Z = \infty$ first. The effect of Z will be examined later. We plot the modulus (absolute value) of the complex displacement at the right end of the rod versus the normalized driving frequency in Fig. 3.18a for two values of $Q = 50$ and 100. These values of Q are relatively small for the material damping of common crystals. We exaggerate the damping a little so that the resonances are not too sharp and are more visible in the figure. In the frequency range shown, there are three resonances with frequencies $\omega^{(n)}$. It can be seen that the values of $\omega^{(n)}/\omega_0^{(1)}$ are close to 1, 3, and 5 where

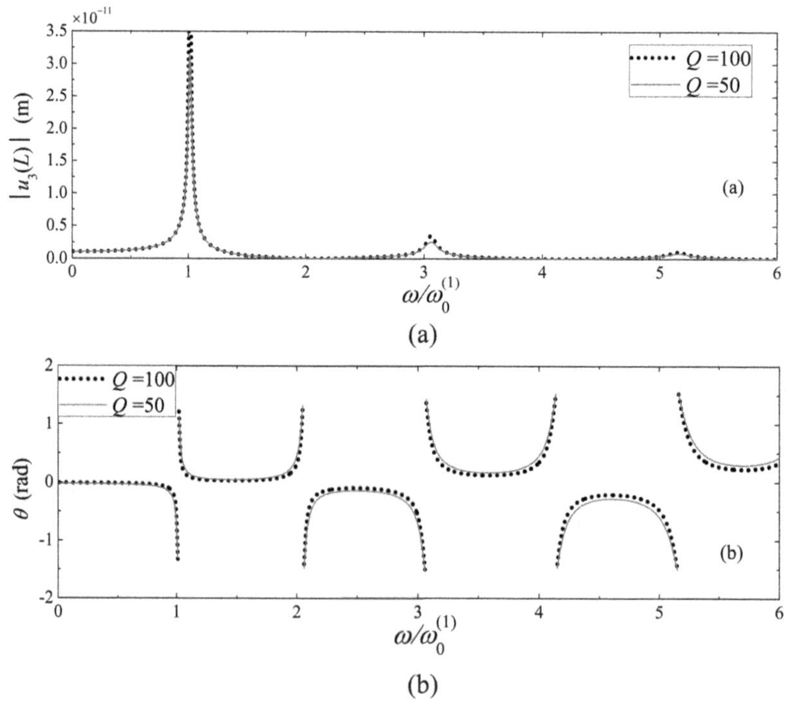

Fig. 3.18 (**a**) $|u_3(L)|$ and (**b**) $\theta = \tan^{-1}[\mathrm{Im}\{u_3(L)\}/\ \mathrm{Re}\ \{u_3(L)\}]$ versus the normalized driving frequency showing resonances. $Z = \infty$

$\omega_0^{(1)} = 1.3198 \times 10^{10}$ rad/s. Below the first resonance, there is a relatively small but finite response which is the static response in the low-frequency limit. The top of the first peak of the curve with $Q = 100$ is 4.79×10^{-11}m which has been cut off in the figure for a better global view. When $Q = 200$, 500, and 5000, the corresponding peak values of the first resonance are 6.35, 7.72 and 8.74×10^{-11}m, respectively, increasing monotonically because of less damping. When Q is very large, the amplitude at the first resonance becomes less sensitive to Q because in this case the dissipation due to semiconduction becomes more influential than material damping. (b) shows the phase angle defined by $\theta = \tan^{-1}[\mathrm{Im}\{u_3(L)\}/\mathrm{Re}\{u_3(L)\}]$. At resonances there is a phase jump of π. The jumps at $\omega/\omega_0^{(1)}$ close to 2 and 4 are caused by $\mathrm{Re}\{u_3(L)\} = 0$ and do not represent resonances. For an elastic rod, the resonance frequencies are exactly odd integral multiples of $\omega_0^{(1)}$. For a piezoelectric dielectric rod, the three frequencies are higher because of the piezoelectric stiffening effect. If the piezoelectric rod is also semiconducting, the mobile charges screen the polarization charges, reduce the piezoelectric stiffening effect, and lower the frequencies. Therefore $\omega^{(n)}$ are somewhat higher than $\omega_0^{(n)}$.

Next we examine the displacement fields at resonances (modes). Strictly speaking, vibration modes are defined by the eigenvalue problem of a free vibration analysis. They can also be felt from our forced vibration analysis by examining the displacement fields at resonances as follows. Based on the complex notation used, the physical field of the mechanical displacement can be obtained from the real part of the complex displacement through

$$\begin{aligned} \mathrm{Re}\{u_3 \exp(i\omega t)\} &= \mathrm{Re}\left\{\left(u_3^R + iu_3^I\right)(\cos\omega t + i\sin\omega t)\right\} \\ &= u_3^R \cos\omega t - u_3^I \sin\omega t, \\ u_3^R &= \mathrm{Re}\{u_3\}, \quad u_3^I = \mathrm{Im}\{u_3\}, \end{aligned} \tag{3.148}$$

which depends on both the real and imaginary parts of the complex displacement. When $Q = 100$, in Fig. 3.19, we plot both the real and imaginary parts of the axial mechanical displacement at the three resonance frequencies. The displacement vanishes at the fixed left end as dictated by the boundary condition. At the first resonance, the displacement does not have nodal points (zeros) except at the fixed left end. At the second and the third resonances, the displacement has one and two nodal points, respectively, in addition to the left end. These are typical behaviors of modes at resonances. Corresponding to the relatively high peak at the first resonance shown in Fig. 3.18a, the displacement at the first resonance is relatively large.

As a mechanical-to-electrical energy converter, the output power and efficiency are of fundamental interest. The mechanical input power P_1 at the right end of the rod and the electrical output power P_2 on Z can be calculated from

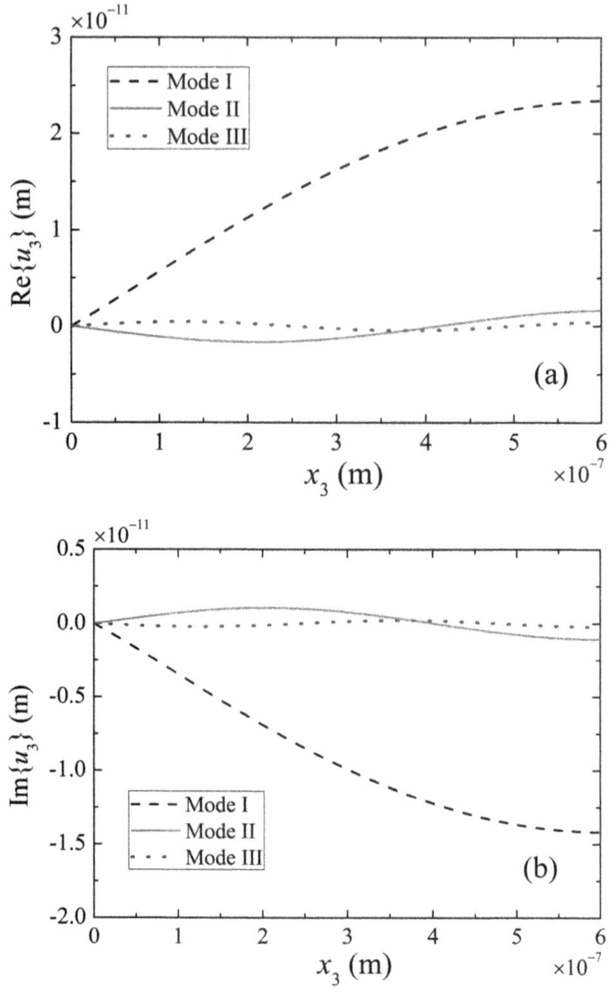

Fig. 3.19 Displacement fields at the first three resonances. $Z=\infty$. $Q=100$. (**a**) Real parts. (**b**) Imaginary parts

$$\begin{aligned} P_1 &= \frac{1}{4}\{F[\dot{u}_3(L)]^* + F^*[\dot{u}_3(L)]\}, \\ P_2 &= \frac{1}{4}\{IV^* + IV^*\}, \end{aligned} \tag{3.149}$$

where an asterisk represents complex conjugate. The efficiency is given by P_2/P_1.

Figure 3.20 shows P_1, P_2, and P_2/P_1 versus the driving frequency of the applied force when Z is real and represents a resistor load. P_1 and P_2 assume maxima at resonance frequencies as expected. As the ratio between two functions of the driving frequency, the behavior of the efficiency is somewhat involved, not necessarily assuming maxima at resonance frequencies.

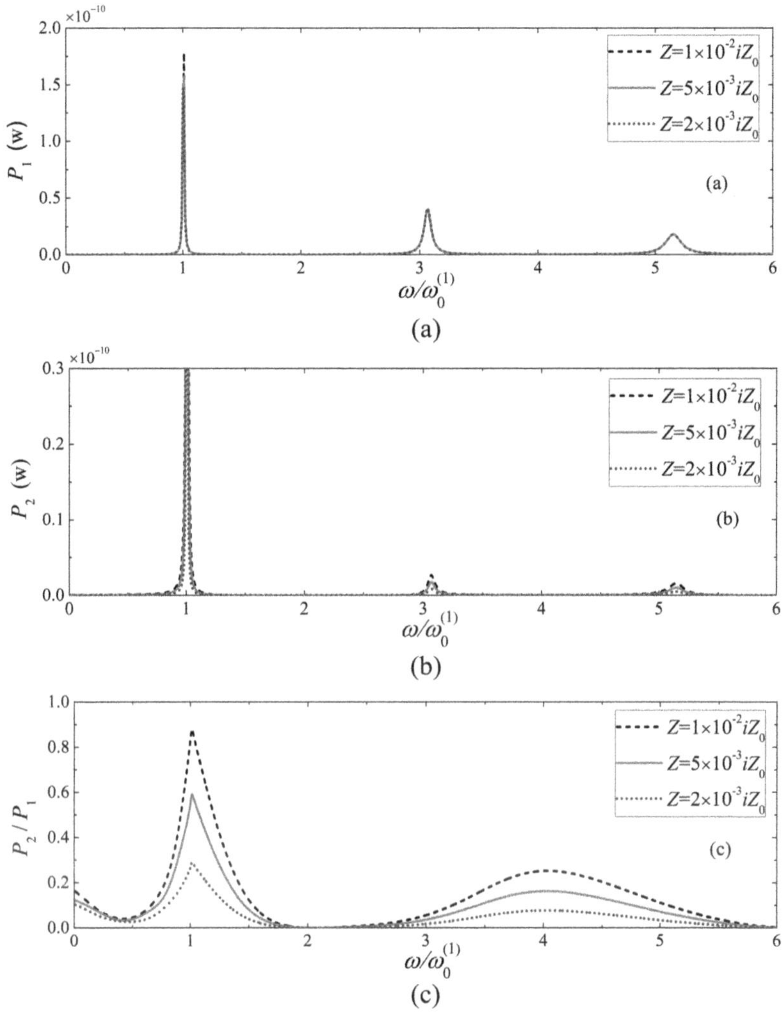

Fig. 3.20 (**a**) Input power verses frequency. (**b**) Output power versus frequency. (**c**) Efficiency versus frequency. $n_0 = 1 \times 10^{22}/\text{m}^3$

3.12 Transient Vibration

Consider the piezoelectric semiconductor rod in Fig. 3.21. The left end is fixed. At $t = 0$, the right end is under the action of a suddenly applied extensional force F which produces an extensional stress $f = F/A$ there [9].

The one-dimensional governing equations are the same as Eqs. (3.131), (3.132), (3.133), (3.134), and (3.135). We consider a ZnO rod doped into an n-type

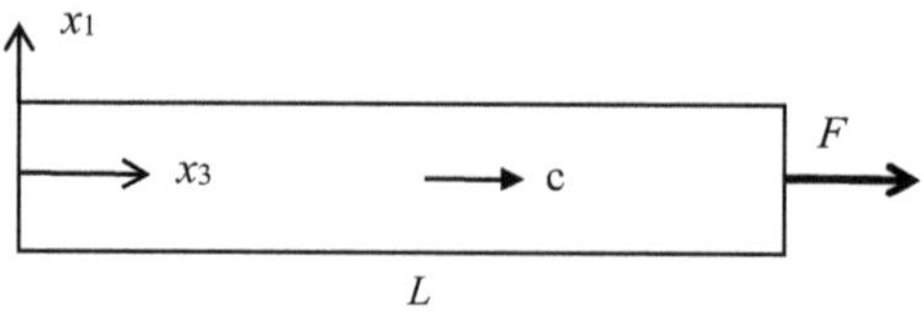

Fig. 3.21 A piezoelectric semiconductor rod under the sudden application of an end force

semiconductor with $p \cong 0$. As an approximation, we neglect the charge on the right-hand side of Eq. $(3.131)_2$. This decouples the problem into two one-way coupled subproblems. One consists of the equations of piezoelectricity for the mechanical displacement u_3 and the electric potential φ. The other is the continuity equation for the electron concentration perturbation Δn. The two subproblems are treated one at a time below. The piezoelectricity subproblem is governed by

$$\frac{\partial T_3}{\partial x_3} = \rho \ddot{u}_3, \quad \frac{\partial D_3}{\partial x_3} = 0. \tag{3.150}$$

Using Eqs. (3.133) and (3.135), we can write Eq. (3.150) as

$$\begin{aligned} \bar{c}_{33} u_{3,33} + \bar{e}_{33} \varphi_{,33} &= \rho \ddot{u}_3, \\ \bar{e}_{33} u_{3,33} - \bar{\varepsilon}_{33} \varphi_{,33} &= 0. \end{aligned} \tag{3.151}$$

The boundary conditions are

$$\begin{aligned} u_3 = 0, \quad D_3 = 0, \quad x_3 = 0, \\ T_3 = f, \quad D_3 = 0, \quad x_3 = L. \end{aligned} \tag{3.152}$$

The initial conditions are

$$u_3 = 0, \quad \dot{u}_3 = 0, \quad \varphi = 0, \quad t = 0. \tag{3.153}$$

From Eq. $(3.151)_2$ and the boundary conditions on the electric displacement in Eq. (3.152), we have

$$D_3 = \bar{e}_{33} u_{3,3} - \bar{\varepsilon}_{33} \varphi_{,3} \equiv 0. \tag{3.154}$$

Then

$$T_3 = \bar{c}_{33} u_{3,3} + \bar{e}_{33} \varphi_{,3} = \bar{c}_{33} u_{3,3} + \bar{e}_{33} \frac{\bar{e}_{33}}{\bar{\varepsilon}_{33}} u_{3,3} = \hat{c}_{33} u_{3,3}, \tag{3.155}$$

where

$$\hat{c}_{33} = \bar{c}_{33} + \frac{\bar{e}_{33}^2}{\bar{\varepsilon}_{33}} \tag{3.156}$$

is a piezoelectrically stiffened elastic constant. Using Eq. $(3.151)_2$ to eliminate the potential, we write Eq. $(3.151)_1$ as

$$\hat{c}_{33}u_{3,33} = \rho \ddot{u}_3. \tag{3.157}$$

The problem for u_3 consists of Eq. (3.157) and the following initial and boundary conditions:

$$\begin{aligned} u_3 &= 0, \quad x_3 = 0, \\ \hat{c}_{33}u_{3,3} &= f, \quad x_3 = L, \end{aligned} \tag{3.158}$$

$$u_3 = 0, \quad \dot{u}_3 = 0, \quad t = 0. \tag{3.159}$$

To homogenize the boundary conditions in Eq. (3.158), we let

$$u_3 = u + \frac{f}{\hat{c}_{33}}x_3. \tag{3.160}$$

The problem for u is

$$\hat{c}_{33}u_{,33} = \rho \ddot{u}, \tag{3.161}$$

$$\begin{aligned} u &= 0, \quad x_3 = 0, \\ \hat{c}_{33}u_{,3} &= 0, \quad x_3 = L, \end{aligned} \tag{3.162}$$

$$u = -\frac{f}{\hat{c}_{33}}x_3, \quad \dot{u} = 0, \quad t = 0. \tag{3.163}$$

Mathematically, Eqs. (3.161), (3.162), and (3.163) are a rather standard problem. Its solution can be obtained in a straightforward manner by the method of separation of variables as

$$u(x_3, t) = \sum_{m=0}^{\infty} A_m \cos\left(\frac{\pi + 2m\pi}{2L}ct\right) \sin\left(\frac{\pi + 2m\pi}{2L}x_3\right), \tag{3.164}$$

where

$$\begin{aligned} A_m &= \frac{2}{L}\int_0^L \left(-\frac{f}{\hat{c}_{33}}\right) x_3 \sin\left(\frac{\pi + 2m\pi}{2L}x_3\right) dx_3, \quad m = 0, 1, 2, \cdots, \\ c &= \sqrt{\frac{\hat{c}_{33}}{\rho}}. \end{aligned} \tag{3.165}$$

Once u is obtained, u_3 and φ are obtained from Eqs. (3.160) and (3.154), respectively. For uniqueness, we let $\varphi = 0$ at $x_3 = 0$. As an example, consider a ZnO rod with $L = 600$ nm and cross-sectional area $A = 2.598 \times 10^{-14}\text{m}^2$. The end force

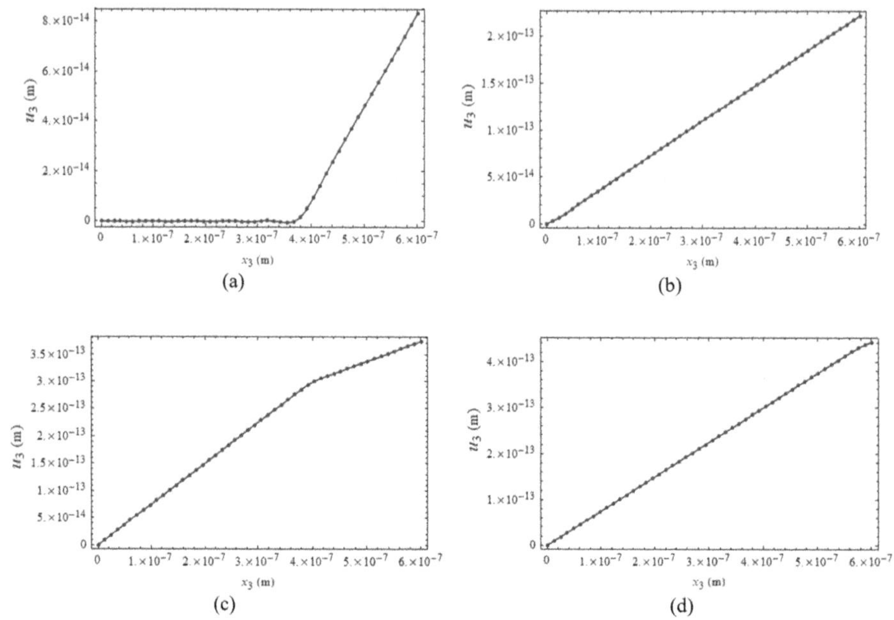

Fig. 3.22 Distribution of u_3 along the rod at different time instants. $m = 0$–15. (**a**) $t = 4 \times 10^{-11}$s. (**b**) $t = 10.7 \times 10^{-11}$s. (**c**) $t = 18 \times 10^{-11}$s. (**d**) $t = 22 \times 10^{-11}$s

$F = 1.7$ nN which produces an end stress of $f = 0.065$ MPa. $n_0 = 10^{23}/\text{m}^3$. The series in Eq. (3.164) converges rapidly. Sixteen terms are used. Figure 3.22 presents a series of snapshots of the distribution of u_3 along the rod at different time instants as the initial disturbance at the right end propagates to the left along the rod, hits the left end and gets reflected there, and then propagates back to the right end. The wave described by Eq. (3.161) is nondispersive and has a finite propagation speed. In (a), the right part of the rod has felt the disturbance with an essentially linearly distributed displacement, but the left part has not felt the disturbance yet. The extensional wave speed can be determined as $c = \sqrt{\hat{c}_{33}/\rho} = 5546\ \text{m/s}$. In our numerical example, we have $L/c = 10.8 \times 10^{-11}$s which is the time for a disturbance to propagate through the entire rod, and it agrees with (b) very well. The propagation of φ is found to be similar to u_3 and is not shown.

The equation for Δn is obtained from Eqs. $(3.132)_2$ and $(3.134)_2$ as

$$\frac{\partial}{\partial t}(\Delta n) = -n_0 \mu_{33}^n \varphi_{,33} + D_{33}^n \frac{\partial^2(\Delta n)}{\partial x_3^2}, \tag{3.166}$$

with the following boundary and initial conditions:

$$J_3 = qn_0\mu_{33}^n E_3 + qD_{33}^n \frac{\partial(\Delta n)}{\partial x_3} = 0, \quad x_3 = 0, L, \tag{3.167}$$

$$\Delta n = 0, \quad t = 0. \tag{3.168}$$

To make the boundary conditions in Eq. (3.167) homogeneous, we let

$$\Delta n = \widehat{n} + \frac{n_0\mu_{33}^n}{D_{33}^n}\varphi. \tag{3.169}$$

The initial-boundary-value problem for $\widehat{n}$ is

$$\frac{\partial \widehat{n}}{\partial t} = D_{33}^n \frac{\partial^2 \widehat{n}}{\partial x_3^2} - \frac{n_0\mu_{33}^n}{D_{33}^n}\frac{\partial \varphi}{\partial t}, \tag{3.170}$$

$$\frac{\partial \widehat{n}}{\partial x_3} = 0, \quad x_3 = 0, L, \tag{3.171}$$

$$\widehat{n} = 0, \quad t = 0. \tag{3.172}$$

Equations (3.170), (3.171), and (3.172) form a standard mathematical problem. Its solution by separation of variables and Laplace transform is

$$\widehat{n}(x_3, t) = \sum_{r=0}^{\infty} \cos\left(\frac{r\pi}{L}x_3\right) \int_0^t a_r(\tau) \exp\left[-\left(\frac{r\pi}{L}\sqrt{D_{33}^n}\right)^2 (t-\tau)\right] d\tau, \tag{3.173}$$

where

$$\begin{aligned} a_0(t) &= \frac{1}{L}\int_0^L \left(-\frac{n_0\mu_{33}^n}{D_{33}^n}\frac{\partial \varphi}{\partial t}\right) dx_3, \quad r = 0, \\ a_r(t) &= \frac{2}{L}\int_0^L \left(-\frac{n_0\mu_{33}^n}{D_{33}^n}\frac{\partial \varphi}{\partial t}\right) \cos\left(\frac{r\pi}{L}x_3\right) dx_3, \quad r = 1, 2, 3 \cdots. \end{aligned} \tag{3.174}$$

From the series solution in Eqs. (3.173) and (3.169), we show in Fig. 3.23 the electron concentration perturbation at the same time instants as those in Fig. 3.22. Fundamentally different from the mechanical disturbance governed by the hyperbolic wave equation in Eq. (3.161) with a finite wave speed, the motion of the electrons is governed by the parabolic equation in Eq. (3.170) with effectively an infinite speed for signal propagation. Therefore, in Fig. 3.23a, while mechanically the left part of the rod is still unperturbed, the electrons along the entire rod have felt the disturbance and been disturbed. Since the rod is electrically isolated, the total number of electrons is conserved. Therefore the integration of Δn over [0,L] has to vanish, which can be qualitatively seen in the figure. The series for Δn converges very rapidly. The results corresponding to seven and eight terms are indistinguishable when plotted in the scale of Fig. 3.23.

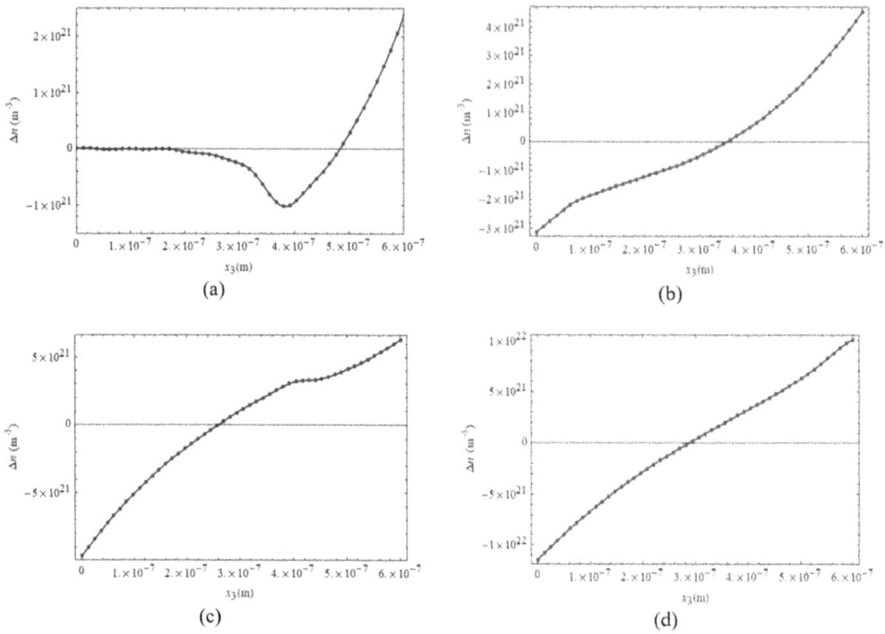

Fig. 3.23 Distribution of Δn along the rod at different time instants. m = 0–15, r = 0–7. (**a**) $t = 4 \times 10^{-11}$s. (**b**) $t = 10.7 \times 10^{-11}$s. (**c**) $t = 18 \times 10^{-11}$s. (**d**) $t = 22 \times 10^{-11}$s

3.13 Nonlinear Waves

Consider the rod shown in Fig. 3.24. It is doped into an n-type semiconductor with $p \cong 0$. We study the propagation of extensional waves with consideration of electrical nonlinearity [10]. The analysis is somewhat crude with several approximations. It is hoped that the solution can still capture some of the physics involved.

The governing equations are as follows:

$$\begin{aligned} T_{3,3} &= \rho \ddot{u}_3, \\ D_{3,3} &= q\left(-n + N_D^+\right), \\ J_{3,3}^n &= q\dot{n}, \end{aligned} \tag{3.175}$$

$$\begin{aligned} T_3 &= \bar{c}_{33} S_3 - \bar{e}_{33} E_3, \\ D_3 &= \bar{e}_{33} S_3 + \bar{\varepsilon}_{33} E_3, \\ J_3^n &= qn\mu_{33} E_3 + qD_{33} n_{,3}, \end{aligned} \tag{3.176}$$

$$\begin{aligned} S_3 &= u_{3,3}, \\ E_3 &= -\varphi_{,3}, \end{aligned} \tag{3.177}$$

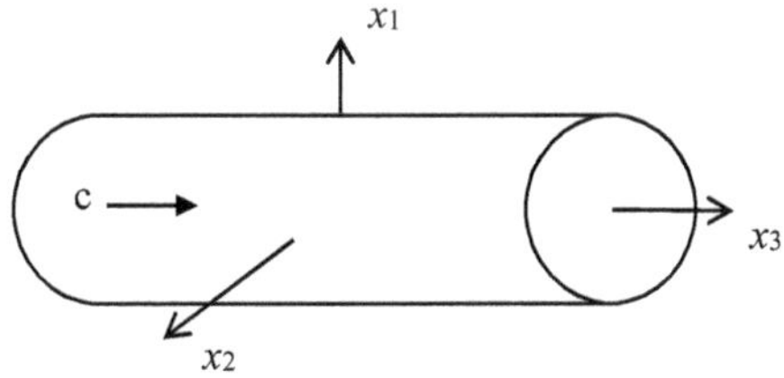

Fig. 3.24 A piezoelectric semiconductor rod of crystals of class (6mm)

where $\mu_{33} = \mu_{33}^n$ and $D_{33} = D_{33}^n$. With the use of Eq. (3.177), we write Eq. (3.176) as

$$\begin{aligned} T_3 &= \bar{c}_{33} u_{3,3} + \bar{e}_{33} \varphi_{,3}, \\ D_3 &= \bar{e}_{33} u_{3,3} - \bar{\varepsilon}_{33} \varphi_{,3}, \\ J_3 &= -qn\mu_{33}\varphi_{,3} + qD_{33} n_{,3}. \end{aligned} \tag{3.178}$$

The substitution of Eq. (3.178) into Eq. (3.175) yields three equations for u_3, φ, and n:

$$\begin{aligned} \bar{c}_{33} u_{3,33} + \bar{e}_{33} \varphi_{,33} &= \rho \ddot{u}_3, \\ \bar{e}_{33} u_{3,33} - \bar{\varepsilon}_{33} \varphi_{,33} &= q(N^+ - n), \\ -\mu_{33}\varphi_{,33} n - \mu_{33}\varphi_{,3} n_{,3} + D_{33} n_{,33} &= \dot{n}. \end{aligned} \tag{3.179}$$

The first two terms in Eq. $(3.179)_3$ are nonlinear.

Assuming low doping and low carrier concentration, we make an approximation by dropping the charges $q(N^+ - n)$ on the right-hand side of Eq. $(3.179)_2$. This reduces Eq. (3.179) into two one-way coupled systems of

$$\begin{aligned} \bar{c}_{33} u_{3,33} + \bar{e}_{33} \varphi_{,33} &= \rho \ddot{u}_3, \\ \bar{e}_{33} u_{3,33} - \bar{\varepsilon}_{33} \varphi_{,33} &= 0, \end{aligned} \tag{3.180}$$

and

$$-\mu_{33}\varphi_{,33} n - \mu_{33}\varphi_{,3} n_{,3} + D_{33} n_{,33} = \dot{n}. \tag{3.181}$$

Equation (3.180) is the classical theory of piezoelectricity within which harmonic extensional waves in the rod are described by

$$u_3 = A\cos\psi, \quad \varphi = B\cos\psi, \tag{3.182}$$

where

$$\psi = \xi x_3 - \omega t, \quad \omega^2 = \frac{\widehat{c}_{33}}{\rho}\xi^2,$$
$$\widehat{c}_{33} = \overline{c}_{33} + \frac{\overline{e}_{33}^2}{\overline{\varepsilon}_{33}}, \quad B = \frac{\overline{e}_{33}}{\overline{\varepsilon}_{33}}A. \tag{3.183}$$

The electric field accompanying the wave in Eq. (3.182) is given by

$$E = B' \sin\psi, \quad B' = \frac{\overline{e}_{33}}{\overline{\varepsilon}_{33}}A\xi. \tag{3.184}$$

With the electric potential in Eq. (3.182) considered as known, Eq. (3.181) becomes linear in n. However, it has variable coefficients depending on both x_3 and t through φ and is still a mathematically challenging problem. We consider uniform doping and denote $\overline{n} = N_D^+$. We write n as the sum of a uniform static part $\overline{n}$ which is a constant and a dynamic part $\widehat{n}(x_3, t)$ caused by the waves in Eq. (3.182), i.e.,

$$n(x_3, t) = \overline{n} + \widehat{n}(x_3, t). \tag{3.185}$$

Substituting Eq. (3.185) into Eq. (3.181), we obtain the following equation for $\widehat{n}$:

$$\dot{\widehat{n}} = \mu_{33}E_{3,3}\overline{n} + \mu_{33}E_{3,3}\widehat{n} + \mu_{33}E_3\widehat{n}_{,3} + D_{33}\widehat{n}_{,33}, \tag{3.186}$$

where the electric field E is given by Eq. (3.184) which, when substituted into Eq. (3.186), results in

$$\dot{\widehat{n}} = \mu_{33}\overline{n}B'\xi\cos\psi + \mu_{33}B'\xi\cos\psi\,\widehat{n} + \mu_{33}B'\sin\psi\,\widehat{n}_{,3} + D_{33}\widehat{n}_{,33}. \tag{3.187}$$

The first three terms on the right-hand side of Eq. (3.187) are from the nonlinear drift current, and the last term is from the diffusion current. The first term does not depend on $\widehat{n}$ and is effectively a driving term from the wave in Eq. (3.182).

Consider a special case of Eq. (3.187) first. If $\widehat{n}$ is very small and only the first term on the right-hand side of Eq. (3.187) is kept, the equation reduces to

$$\dot{\widehat{n}} = \mu_{33}\overline{n}B'\xi\cos\psi, \tag{3.188}$$

which admits the following solution

$$\widehat{n} = -\mu_{33}\overline{n}B'\frac{\xi}{\omega}\sin\psi. \tag{3.189}$$

Equation (3.189) describes a simple sinusoidal wave motion of the carriers driven by the electromechanical wave in Eq. (3.182).

In general, when all of the terms in Eq. (3.187) are considered, we write the solution using the following trigonometric series:

$$\hat{n} = \sum_{m=1}^{M} a_m \cos m\psi + b_m \sin m\psi, \qquad (3.190)$$

where a_m and b_m are undetermined constants. We substitute Eq. (3.190) into Eq. (3.187), write both sides of the resulting equation in terms of $\cos m\psi$ and $\sin m\psi$ using trigonometric identities, and set their corresponding coefficients equal. This yields a hierarchy of linear equations for a_m and b_m. These equations are truncated and solved on a computer.

As a numerical example, consider a rod with a cross-sectional dimension characterized by some diameter $d = 1$ mm. $\bar{n} = 10^{13}/\text{m}^3$. For a wave with wavelength $\lambda = 10$ mm which is much larger than the cross-sectional dimension, the wave number $\xi = 2\pi/\lambda = 628.3$ 1/m. Then, from Eq. (3.183), the wave frequency is $\omega = 3.49 \times 10^6$ 1/s. The wave amplitude A will be varied in the computation. In the case when $A = 3 \times 10^{-8}$m, the corresponding amplitude of the axial strain $S_3 = u_{3,3}$ is $A\xi = 18.85 \times 10^{-6}$. We calculate $\hat{n}$ for increasing values of A in the range from 10^{-9}m to 3×10^{-8}m with five terms in the series in Eq. (3.190), i.e., $M = 5$ which has been verified numerically to have sufficient accuracy for examining the qualitative behaviors of the waves. The results are summarized in Fig. 3.25. In (a), when A is small, the wave is essentially sinusoidal. As A increases, deviation from a sinusoidal wave becomes visible. The height of the crests is larger than the depth of the troughs, and the crests are narrower than the troughs. As A increases further in (b), the wave deviates from a sinusoidal one severely. These are believed to be due to the electrical nonlinearity and are to be verified numerically by, e.g., COMSOL.

3.14 Extension of a PN Junction

In this section we study the extension of a piezoelectric semiconductor rod with a PN junction (see Fig. 3.26). The rod has a unit cross-sectional area and is under an axial force F. It is electrically open at both ends. Both homogeneous and heterogeneous junctions will be considered. For a homogeneous junction, the material is uniform along the entire rod, but the two halves are doped differently. The left half is dominated by holes and the right half by electrons. For a heterogeneous junction, the materials of the two halves are different. We consider the case of a relatively simple heterogeneous junction obtained by reversing the c-axis of the material of the left half. When the c-axis is reversed, the relevant piezoelectric constants change signs. The problem is static. Below is a combination of a linear theoretical analysis and a nonlinear numerical analysis by COMSOL [11]. A nonlinear analysis using both the perturbation method and COMSOL can be found in [12].

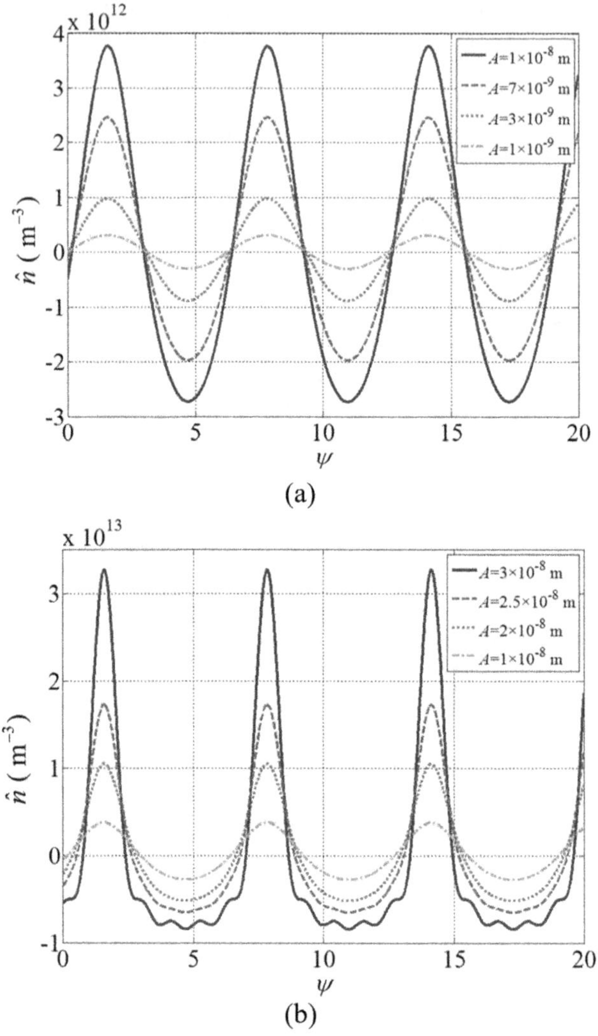

Fig. 3.25 Effect of the wave amplitude A

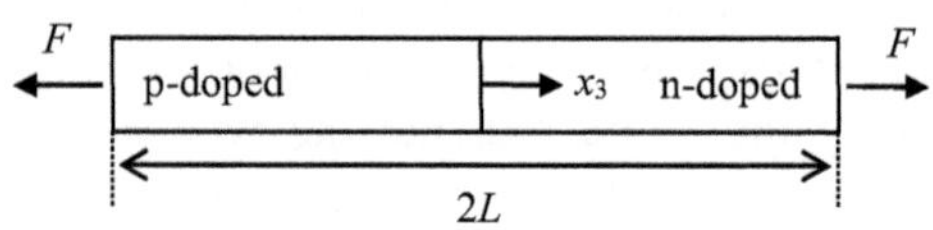

Fig. 3.26 A PN junction in a piezoelectric semiconductor rod in extension

The one-dimensional governing equations are

$$\begin{aligned} T_{33,3} &= \rho \ddot{u}_3, \\ D_{3,3} &= q(\Delta p - \Delta n), \\ J^p_{3,3} &= -q\dot{p}, \\ J^n_{3,3} &= q\dot{n}, \end{aligned} \tag{3.191}$$

$$\begin{aligned} T_{33,3} &= \overline{c}_{33}S_{33} - \overline{e}_{33}E_3, \\ D_3 &= \overline{e}_{33}S_{33} + \overline{\varepsilon}_{33}E_3, \end{aligned} \tag{3.192}$$

$$\begin{aligned} J^p_3 &= qp\mu^p_{33}E_3 - qD^p_3 p_{,3} \\ &\cong qp_0\mu^p_{33}E_3 - qD^p_3(\Delta p)_{,3}, \\ J^n_3 &= qp\mu^n_{33}E_3 + qD^n_3 n_{,3} \\ &\cong qn_0\mu^n_{33}E_3 + qD^n_3(\Delta n)_{,3}, \end{aligned} \tag{3.193}$$

$$S_{33} = u_{3,3}, \quad E_3 = -\varphi_{,3}. \tag{3.194}$$

The substituon of Eqs. (3.192), (3.193) and (3.194) into Eq. (3.191) yields four second-order ordinary differential equations for u_3, φ, Δp, and Δn. The equations may be linear or nonlinear depending on whether the linearized or nonlinear version of Eq. (3.193) is used. The axial electric polarization P_3 and the effective polarization charge density ρ^P can be calculated from the following expressions:

$$\begin{aligned} P_3 &= D_3 - \varepsilon_0 E_3, \\ \rho^P &= -P_{3,3}. \end{aligned} \tag{3.195}$$

We follow the simplified notation in Eq. (3.80). The boundary and continuity conditions are

$$\begin{aligned} T(-L) = F, \quad D(-L) = 0, \\ J^p(-L) = 0, \quad J^n(-L) = 0, \end{aligned} \tag{3.196}$$

$$\begin{aligned} u(0^-) = u(0^+), \quad T(0^-) = T(0^+), \\ \varphi(0^-) = \varphi(0^+), \quad D(0^-) = D(0^+), \end{aligned} \tag{3.197}$$

$$\begin{aligned} p(0^-) = p(0^+), \quad J^p(0^-) = J^p(0^+), \\ n(0^-) = n(0^+), \quad J^n(0^-) = J^n(0^+), \end{aligned} \tag{3.198}$$

$$\begin{aligned} T(L) = F, \quad D(L) = 0, \\ J^p(L) = 0, \quad J^n(L) = 0. \end{aligned} \tag{3.199}$$

Δp and Δn satisfy the following global charge neutrality conditions:

$$\int_{-L}^{L} \Delta p dx = 0, \quad \int_{-L}^{L} \Delta n dx = 0. \tag{3.200}$$

Only one of Eq. (3.200) is independent. To determine the displacement and potential fields uniquely, we choose a reference point, e.g., $x = a$, and set

$$u(a) = 0, \quad \varphi(a) = 0. \tag{3.201}$$

Mathematically, we need to find solutions in the two regions in $-L<x<0$ and $0<x<L$ separately and apply boundary and continuity conditions.

For the case of a small F and small carrier concentration perturbations, we perform a theoretical analysis using the linearized constitutive relations for currents in Eq. (3.193). In each region we have a system of linear ordinary differential equations with constant coefficients. The procedure for finding a general solution is straightforward. The junction may be heterogeneous. We use a prime for the material parameters of the left half and a double prime for those of the right half. The general solution for $-L<x<0$ is

$$\Delta p - \Delta n = A_1 \sinh k'(x - L) + B_1 \sinh k'(x + L), \tag{3.202}$$

$$\varphi = -\frac{q}{(k')^2 \varepsilon'^T_{33}} [A_1 \sinh k'(x - L) + B_1 \sinh k'(x + L)] + C_1 x + C_2, \tag{3.203}$$

$$u = \frac{e'}{c'} \frac{q}{(k')^2 \varepsilon'^T_{33}} [A_1 \sinh k'(x - L) + B_1 \sinh k'(x + L)] + C_3 x + C_4, \tag{3.204}$$

$$\Delta p = \frac{p'_0 \mu'^p}{D'^p} \frac{q}{(k')^2 \varepsilon'^T_{33}} [A_1 \sinh k'(x - L) + B_1 \sinh k'(x + L)] + C_5 x + C_6, \tag{3.205}$$

where A_1, B_1, and C_1–C_6 are undetermined constants, and

$$(k')^2 = \left(\frac{p'_0 \mu'^p}{D'^p} + \frac{n'_0 \mu'^n}{D'^n} \right) \frac{q}{\varepsilon'^T_{33}}. \tag{3.206}$$

Similarly, for $0<x<L$, we have

$$\Delta p - \Delta n = A_2 \sinh k''(x - L) + B_2 \sinh k''(x + L), \tag{3.207}$$

$$\varphi = -\frac{q}{(k'')^2 \varepsilon''^T_{33}} [A_1 \sinh k''(x - L) + B_1 \sinh k''(x + L)] + C_7 x + C_8, \tag{3.208}$$

$$u = \frac{e''}{c''} \frac{q}{(k'')^2 \varepsilon''^T_{33}} [A_2 \sinh k''(x - L) + B_2 \sinh k''(x + L)] + C_9 x + C_{10}, \tag{3.209}$$

$$\Delta p = \frac{p_0''\mu''^p}{D''^p}\frac{q}{(k'')^2\varepsilon''^T_{33}}[A_2\sinh k''(x-L) + B_2\sinh k''(x+L)] + C_{11}x$$
$$+ C_{12}, \tag{3.210}$$

where A_2, B_2, and C_7–C_{12} are undetermined constants, and

$$(k'')^2 = \left(\frac{p_0''\mu''^p}{D''^p} + \frac{n_0''\mu''^n}{D''^n}\right)\frac{q}{\varepsilon''^T_{33}}. \tag{3.211}$$

Equations (3.202), (3.203), (3.204), (3.205) and Eqs. (3.207), (3.208), (3.209), (3.210) are subsituted into Eqs. (3.196), (3.197), (3.198), (3.199), (3.200), and (3.201). This results in a system of linear equations for the undetermined constants. The equations are solved on a computer.

As an example, consider a long ZnO fiber with $2L = 30$ μm, $p_0' = 10^{21}$ m^{-3}, $n_0' = 7\times10^{20}$ m^{-3}, $n_0'' = 10^{21}$ m^{-3}, $p_0'' = 7\times10^{20}$ m^{-3}, and $a = -1$ μm. We organize the numerical results below according to increasing magnitude of F. For small values of F, the linear analytical solution and the nonlinear COMSOL numerical solution produce the same result. For medium F, the two solutions begin to show differences. For large F, the linear analytical solution is invalid.

In the case of a small $F = 0$–0.2 MPa, since the linear analytical solution and the nonlinear COMSOL solution agree well, only the analytical solution is presented below. The polarization fields are shown in Fig. 3.27. Away from the junction and on its both sides, the P in (a) for the homogeneous junction has the same value but, that in (b) for the heterogeneous has opposite signs.

The different behaviors of P in Fig. 3.27 have implications on the effective polarization charge density ρ^P they produce according to Eq. $(3.195)_2$, which are shown in Fig. 3.28. For the homogeneous junction in (a), F has no effect, and there are no net polarization charges near the junction. For the heterogeneous junction in (b), F produces negative net polarization charges.

In addition to the distributed effective polarization charge density descibed by ρ^P, there also exists a surface effective polarization charge density σ^P on the interface of the PN junction due to the discontinuity of P there. σ^P can be calculated from

$$\begin{aligned} &\sigma^P = P^- - P^+, \\ &P^- = P(0^-),\ \ P^+ = P(0^+). \end{aligned} \tag{3.212}$$

We list σ^P at the junction interface in Table 3.1. For the homogeneous junction, we always have $\sigma^P \cong 0$. However, for the heterogeneous junction, the magnitude of σ^P increases with F. σ^P is negative and contributes further to the negative net polarization charges near the junction produced by ρ^P in Fig. 3.28b.

For a small $F = 0$–0.2 MPa, Fig. 3.29 shows the effects of F on the built-in potential and electric field near the heterogeneous junction. For the homogeneous junction F has no visible effects. Only the central part of the rod within $|x|<1$ μm is

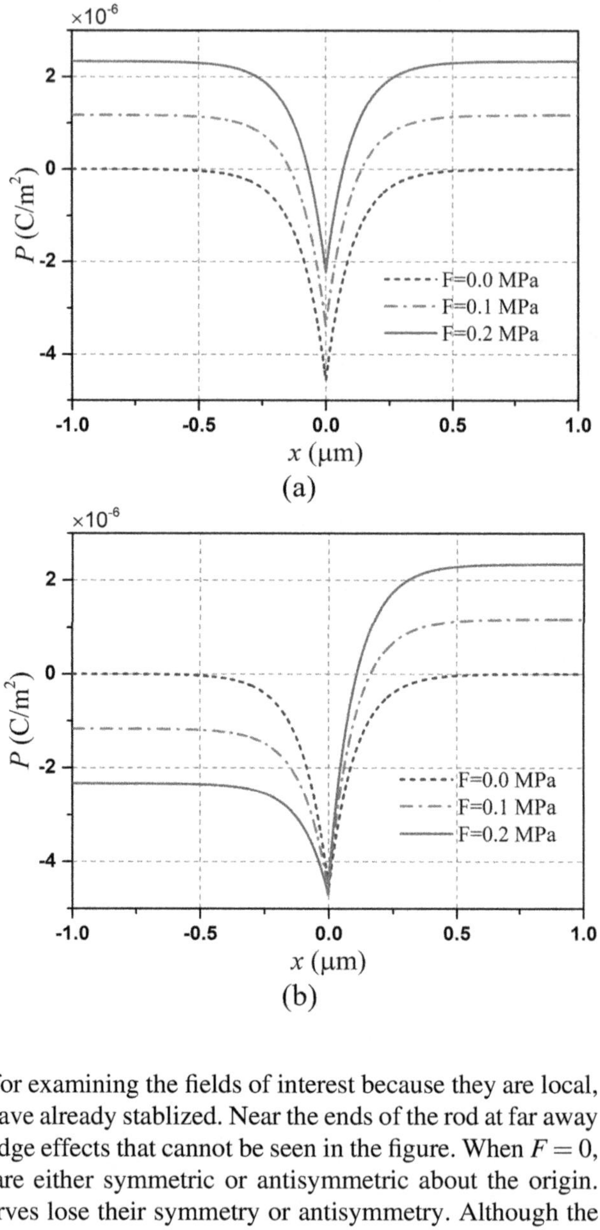

Fig. 3.27 Effects of a small $F = 0$–0.2 MPa on polarization. (**a**) Homogeneous junction. (**b**) Heterogeneous junction

shown, which is sufficient for examining the fields of interest because they are local, and near $x = \pm 1$ μm, they have already stablized. Near the ends of the rod at far away where $x = \pm L$, there exist edge effects that cannot be seen in the figure. When $F = 0$, the corresponding curves are either symmetric or antisymmetric about the origin. When F is nonzero, the curves lose their symmetry or antisymmetry. Although the built-in voltage determined by the potential difference at $x = \pm 1$ μm in (a) is essentially unaffected by F, the slope of the potential field or the built-in electric field in (b) is affected. When $F = 0.2$ MPa, the potential in (a) is continuous but nonsmooth at the origin. This is related to that the corresponding electric field in (b) has a jump dicontinuity at the origin which has been connected by a vertical line in the figure.

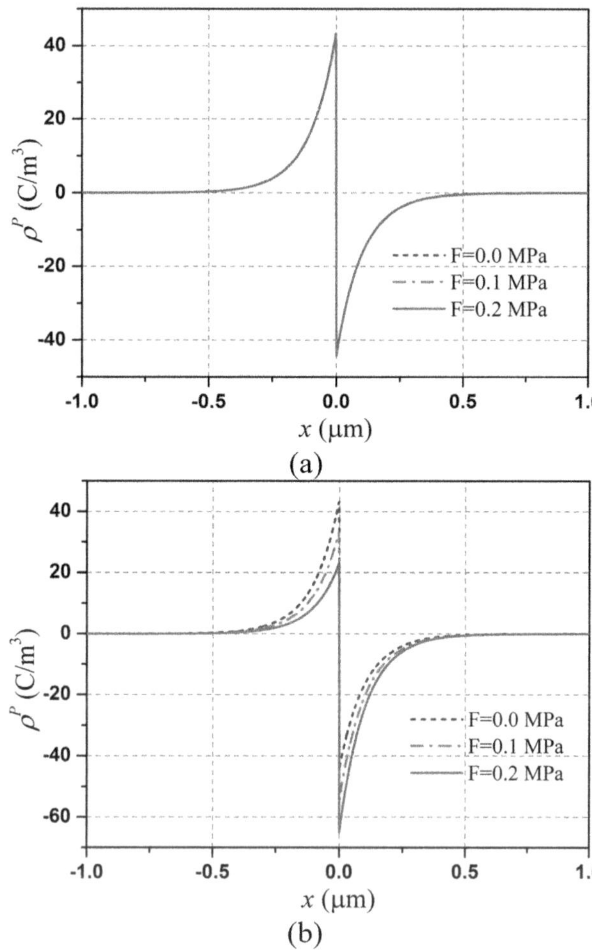

Fig. 3.28 Effects of a small $F = 0$–0.2 MPa on effective polarization charge density. (**a**) Homogeneous junction. (**b**) Heterogeneous junction

Table 3.1 Concentrated effective polarization charge at the junction

Charge (C/m^2)	P^-	P^+	σ^P
Load	$F = 0$ MPa		
Homogenous	-4.5638×10^{-6}	-4.5638×10^{-6}	0
Heterogeneous	-4.5638×10^{-6}	-4.5638×10^{-6}	0
Load	$F = 0.1$ MPa		
Homogeneous	-3.3952×10^{-6}	-3.3952×10^{-6}	0
Heterogeneous	-4.6562×10^{-6}	-4.4714×10^{-6}	-1.8475×10^{-7}
Load	$F = 0.2$ MPa		
Homogeneous	-2.2267×10^{-7}	-2.2267×10^{-7}	0
Heterogeneous	-4.7485×10^{-6}	-4.3790×10^{-6}	-3.6951×10^{-7}

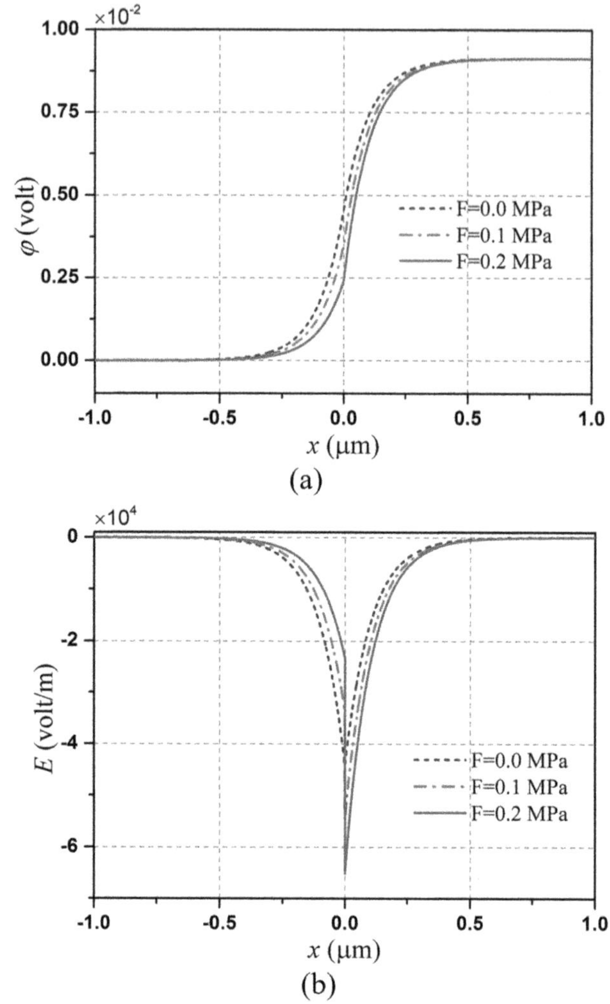

Fig. 3.29 Effects of a small $F = 0$–0.2 MPa in the heterogeneous junction. (**a**) Electric potential φ. (**b**) Electric field E

If we increase the axial force to the medium range of $F = 0.1$–1.0 MPa, the linear analytical and nonlinear numerical solutions begin to show small but visible differences [11]. Only the numerical solution by COMSOL is shown below. The homogeneous junction still does not show any visible effects of F. Therefore only the heterogeneous junction is shown in Fig. 3.30. The potential in (a) is no longer monotonic, and a potential well begins to form. (b) shows that there are more net polarization charges near the junction.

If we increase the axial force further to a large $F = 3$–7 MPa, the differences between the linear analytical and nonlinear numerical solutions become significant, and the linear solution is no longer reliable. Figure 3.31 shows the nonlinear numerical results for the heterogeneous junction only. (a) shows a deep potential

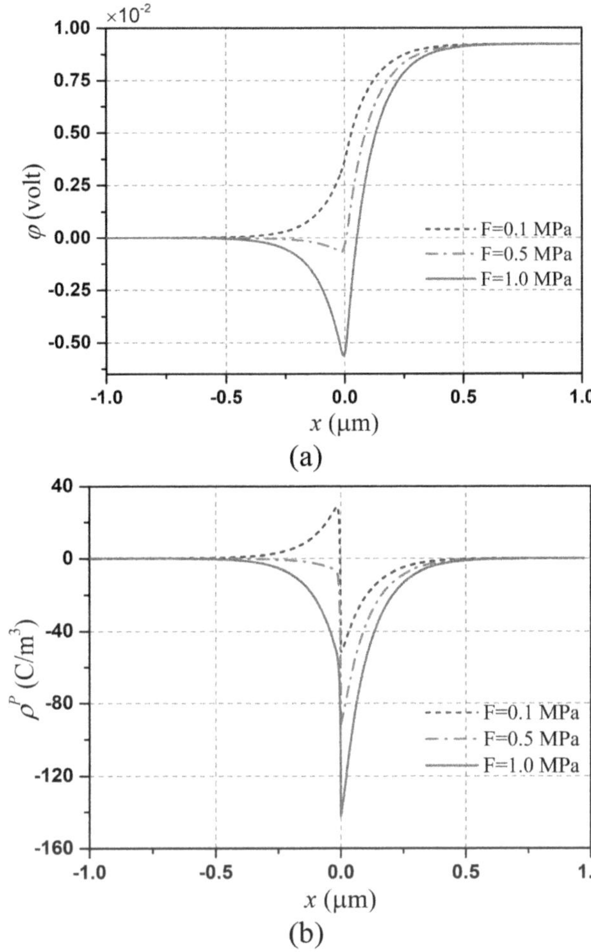

Fig. 3.30 Effects of a medium $F = 0.1$–1.0 MPa in the heterogeneous junction. (**a**) φ by COMSOL. (**b**) ρ^P by COMSOL

well which is related to that the ρ^P in (b) provides more negative polarization charges at the junction.

For $F = 1$–7 MPa, the comparisons of the interface effective polarization charges calculated from the linear analytical and nonlinear numerical solutions are shown in Table 3.2. As F increases, the difference between the two solutions increases as expected.

We now consider the case when the two ends of the rod at $x = \pm L$ are under an applied voltage $2V$ with the following boundary conditions:

$$\varphi(-L) = V, \quad \varphi(L) = -V. \tag{3.213}$$

Our main interest is the current-voltage relation or $I-V$ curve. In view of the well-known nonlinear behavior associated with the typical $I-V$ curves of a PN junction,

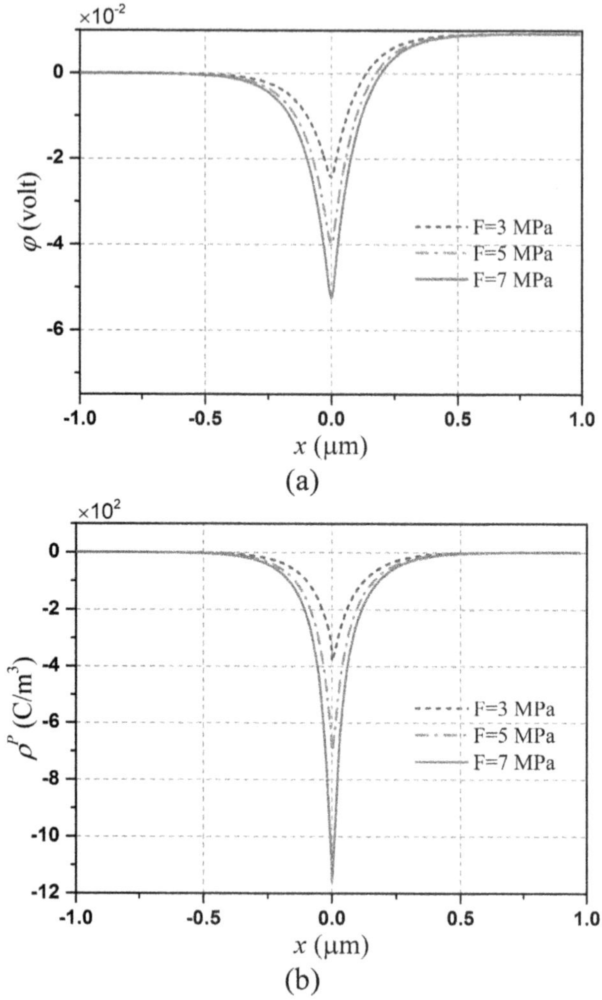

Fig. 3.31 Effects of a large $F = 3$–7 MPa in the heterogeneous junction. (**a**) φ by COMSOL. (**b**) ρ^P by COMSOL

Table 3.2 Effects of F on σ^P in the heterogeneous junction

Load (MPa)		$F = 1$	$F = 3$	$F = 5$	$F = 7$
Analytical solution					
Charge (μC/m^2)	P^-	−5.4875	−7.3351	−9.1826	−11.030
	P^+	−3.6400	−1.7925	0.05509	1.9026
	σ^P	−1.8475	−5.5426	−9.2377	−12.933
COMSOL					
Charge (μC/m^2)	P^-	−5.4904	−7.1446	−8.7423	−10.4040
	P^+	−3.6257	−1.5200	0.7291	3.0498
	σ^P	−1.8647	−5.6246	−9.4714	−13.4538

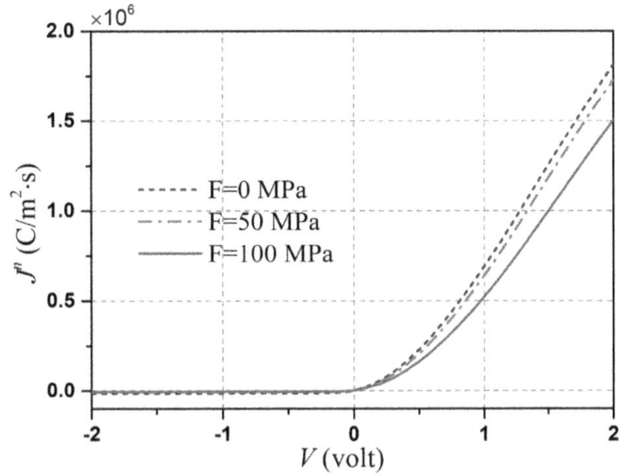

Fig. 3.32 Effects of F on the $I-V$ curves of the heterogeneous junction

for the very large values of F used, the nonlinear version of the constitutive relations for currents is chosen, and a numerical analysis using COMSOL is performed. The results are shown in Fig. 3.32. Again, the homogeneous junction is insensitive to F and is not shown, but the heterogeneous junction shows a stress sensitivity which may be used for piezotronic applications.

3.15 Small Fields Superposed on a Finite Bias

Consider the piezoelectric semiconductor rod shown in Fig. 3.33. We study small dynamic fields superposed on finite static fields called biasing fields in extensional motions of the rod.

The one-dimensional equations with consideration of electrical nonlinearity are

$$\begin{aligned} \frac{\partial T_3}{\partial x_3} &= \rho \ddot{u}_3, \\ \frac{\partial D_3}{\partial x_3} &= q\left(p - n + N_D^+ - N_A^-\right) \ , \end{aligned} \tag{3.214}$$

$$q\dot{p} = -\frac{\partial J_3^p}{\partial x_3}, \quad q\dot{n} = \frac{\partial J_3^n}{\partial x_3}, \tag{3.215}$$

$$\begin{aligned} T_3 &= \overline{c}_{33} S_3 - \overline{e}_{33} E_3, \\ D_3 &= \overline{e}_{33} S_3 + \overline{\varepsilon}_{33} E_3, \end{aligned} \tag{3.216}$$

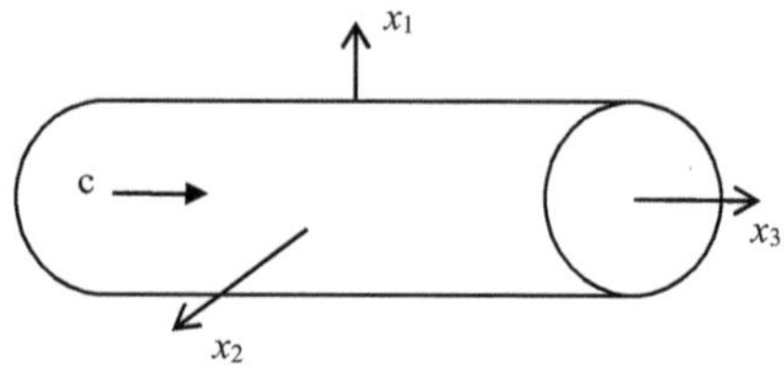

Fig. 3.33 A piezoelectric semiconductor rod of crystals of class (6mm)

$$\begin{aligned} J_3^p &= qp\mu_{33}^p E_3 - qD_{33}^p \frac{\partial p}{\partial x_3}, \\ J_3^n &= qn\mu_{33}^n E_3 + qD_{33}^n \frac{\partial n}{\partial x_3}, \end{aligned} \tag{3.217}$$

$$S_3 = \frac{\partial u_3}{\partial x_3}, \quad E_3 = -\frac{\partial \varphi}{\partial x_3}. \tag{3.218}$$

Let

$$\begin{aligned} u_3(x_3,t) &= \overline{u}(x_3) + \widetilde{u}(x_3,t), \quad \varphi(x_3,t) = \overline{\varphi}(x_3) + \widetilde{\varphi}(x_3,t), \\ S_3(x_3,t) &= \overline{S}(x_3) + \widetilde{S}(x_3,t), \quad E_3(x_3,t) = \overline{E}(x_3) + \widetilde{E}(x_3,t), \\ T_3(x_3,t) &= \overline{T}(x_3) + \widetilde{T}(x_3,t), \quad D_3(x_3,t) = \overline{D}(x_3) + \widetilde{D}(x_3,t), \end{aligned} \tag{3.219}$$

$$\begin{aligned} p(x_3,t) &= \overline{p}(x_3) + \widetilde{p}(x_3,t), \quad n(x_3,t) = \overline{n}(x_3) + \widetilde{n}(x_3,t), \\ J_3^p(x_3,t) &= \overline{J}^p(x_3) + \widetilde{J}^p(x_3,t), \quad J_3^n(x_3,t) = \overline{J}^n(x_3) + \widetilde{J}^n(x_3,t), \end{aligned} \tag{3.220}$$

where we have used an over bar to describe the finite static initial fields and a tilde for the small dynamic incremental fields. For one-dimensional problems of the extension of a rod, to simplify the notation, we denote

$$\begin{aligned} & x_3 = x, \\ & \overline{c}_{33} = c, \quad \overline{e}_{33} = e, \quad \overline{\varepsilon}_{33} = \varepsilon, \\ & \mu_{33}^p = \mu^p, \quad D_{33}^p = D^p, \\ & \mu_{33}^n = \mu^n, \quad D_{33}^n = D^n. \end{aligned} \tag{3.221}$$

Substituting Eqs. (3.219), (3.220), and (3.221) into Eqs. (3.214), (3.215), (3.216), (3.217), and (3.218), collecting zero- and first-order terms of the small incremental fields, we obtain two sets of equations. One is for the static initial fields:

$$\begin{aligned} & \frac{\partial \overline{T}}{\partial x} = 0, \\ & \frac{\partial \overline{D}}{\partial x} = q\left(\overline{p} - \overline{n} + N_D^+ - N_A^-\right), \end{aligned} \tag{3.222}$$

$$0 = -\frac{\partial \overline{J}^p}{\partial x}, \quad 0 = \frac{\partial \overline{J}^n}{\partial x}, \tag{3.223}$$

$$\begin{aligned} \overline{T} &= c\overline{S} - e\overline{E}, \\ \overline{D} &= e\overline{S} + \varepsilon\overline{E}, \end{aligned} \tag{3.224}$$

$$\begin{aligned} \overline{J}^p &= q\overline{p}\mu^p\overline{E} - qD^p\frac{\partial \overline{p}}{\partial x}, \\ \overline{J}^n &= q\overline{n}\mu^n\overline{E} + qD^n\frac{\partial \overline{n}}{\partial x}, \end{aligned} \tag{3.225}$$

$$\overline{S} = \frac{\partial \overline{u}}{\partial x}, \quad \overline{E} = -\frac{\partial \overline{\varphi}}{\partial x}. \tag{3.226}$$

The other is for the small incremental fields:

$$\begin{aligned} \frac{\partial \widetilde{T}}{\partial x} &= \rho\ddot{\widetilde{u}}, \\ \frac{\partial \widetilde{D}}{\partial x} &= q(\widetilde{p} - \widetilde{n}) \ , \end{aligned} \tag{3.227}$$

$$q\dot{\widetilde{p}} = -\frac{\partial \widetilde{J}^p}{\partial x}, \quad q\dot{\widetilde{n}} = \frac{\partial \widetilde{J}^n}{\partial x}, \tag{3.228}$$

$$\begin{aligned} \widetilde{T} &= c\widetilde{S} - e\widetilde{E}, \\ \widetilde{D} &= e\widetilde{S} + \varepsilon\widetilde{E}, \end{aligned} \tag{3.229}$$

$$\begin{aligned} \widetilde{J}^p &= q\overline{p}\mu^p\widetilde{E} + q\mu^p\overline{E}\widetilde{p} - qD^p\frac{\partial \widetilde{p}}{\partial x}, \\ \widetilde{J}^n &= q\overline{n}\mu^n\widetilde{E} + q\mu^n\overline{E}\widetilde{n} + qD^n\frac{\partial \widetilde{n}}{\partial x}, \end{aligned} \tag{3.230}$$

$$\widetilde{S} = \frac{\partial \widetilde{u}}{\partial x}, \quad \widetilde{E} = -\frac{\partial \widetilde{\varphi}}{\partial x}. \tag{3.231}$$

Equations (3.227), (3.228), (3.229), (3.230), and (3.231) are linear in the small incremental fields. Equation (3.230) shows that the initial fields appear as coefficients of the equations for the incremental fields. Since the nonlinearity is due to the drift current, only $\overline{E}$, $\overline{p}$, and $\overline{n}$ affect the incremental fields directly through Eq. (3.230).

References

1. Z.L. Wang, *Piezotronics and Piezo-Phototronics* (Science Press, Beijing, 2012, in Chinese)
2. G.Y. Yang, J.K. Du, J. Wang, J.S. Yang, Electromechanical fields in a nonuniform piezoelectric semiconductor rod. J. Mech. Mater. Struct. **13**, 103–120 (2018)

3. G.Y. Yang, J.K. Du, J. Wang, J.S. Yang, Electromechanical fields in PN junctions with continuously graded doping in piezoelectric semiconductor rods. Under review
4. C.L. Zhang, X.Y. Wang, W.Q. Chen, J.S. Yang, An analysis of the extension of a ZnO piezoelectric semiconductor nanofiber under an axial force. Smart Mater. Struct. **26**, 025030 (2017)
5. G.Y. Yang, J.K. Du, J. Wang, J.S. Yang, Extension of a piezoelectric semiconductor fiber with consideration of electrical nonlinearity. Acta Mech. **229**, 4663–4676 (2018)
6. S.Q. Fan, Y.T. Hu, J.S. Yang, Stress-induced potential barriers and charge distributions in a piezoelectric semiconductor nanofiber. Appl. Math. Mech. (English Ed.) **40**, 591–600 (2019)
7. H.Y. Huang, Z.H. Qian, J.S. Yang, I-V characteristics of a piezoelectric semiconductor nanofiber under local tensile/compressive stress. J. Appl. Phys. **126**, 164902 (2019)
8. G.L. Wang, J.X. Liu, X.L. Liu, W.J. Feng, J.S. Yang, Extensional vibration characteristics and screening of polarization charges in a ZnO piezoelectric semiconductor nanofiber. J. Appl. Phys. **124**, 094502 (2018)
9. W.L. Yang, Y.T. Hu, J.S. Yang, Transient extensional vibration in a ZnO piezoelectric semiconductor nanofiber under a suddenly applied end force. Mater. Res. Express **6**, 025902 (2019)
10. C.L. Zhang, X.Y. Wang, W.Q. Chen, J.S. Yang, Propagation of extensional waves in a piezoelectric semiconductor rod. AIP Adv. **6**, 045301 (2016)
11. R.R. Cheng, C.L. Zhang, W.Q. Chen, J.S. Yang, On the effective polarization charges in the extension of a piezoelectric semiconductor fiber with a PN junction. In Proc. 14th Symp. on Piezoelectricity, Acoustic Waves and Device Applications (SPAWDA), (IEEE, New York, 2019), pp. 1–7
12. M.K. Guo, Y. Li, G.S. Qin, M.H. Zhao, Nonlinear solutions of PN junctions of piezoelectric semiconductors. Acta Mech. **230**, 1825–1841 (2019)

Chapter 4
Bending of Beams

Piezoelectric semiconductor beams or fibers of ZnO in bending or flexural deformation are widely used in devices [1]. Qualitatively, the bending stiffness of a beam is smaller than its extensional stiffness. Therefore bending is suitable in low-frequency applications. Theoretical modeling of bending is more complicated than that of extension. This chapter begins with the establishment of a one-dimensional model for the bending of piezoelectric semiconductor beams. For bending, shear deformation may be important or not in different situations. We derive beams equations for bending with shear deformation first, and then reduce them to bending without shear deformation. For a complete treatment, while deriving the one-dimensional equations for bending, we also include extension because bending and extension may be coupled due to anisotropy or nonlinearity. A few static and dynamic problems are analyzed using the equations derived.

4.1 One-Dimensional Equations for Bending

Consider a piezoelectric semiconductor beam of crystal class (6mm) such as ZnO as shown in Fig. 4.1. The lateral surface is unelectroded and is free from any electromechanical loading. The electric field in the surrounding free space is neglected as usual. These apply to the rest of the chapter. We consider bending in the y-z plane. The cross section and the load are symmetric about the y axis.

For linear bending with small deformation and weak fields the governing equations are Eqs. $(1.1)_1$, (1.10), (1.3), (1.9) with uniform doping, and (1.5):

$$\begin{aligned} T_{ji,j} &= \rho \ddot{u}_i, \\ D_{i,i} &= q(\Delta p - \Delta n) \end{aligned} \quad , \tag{4.1}$$

J. Yang, *Analysis of Piezoelectric Semiconductor Structures*,
https://doi.org/10.1007/978-3-030-48206-0_4

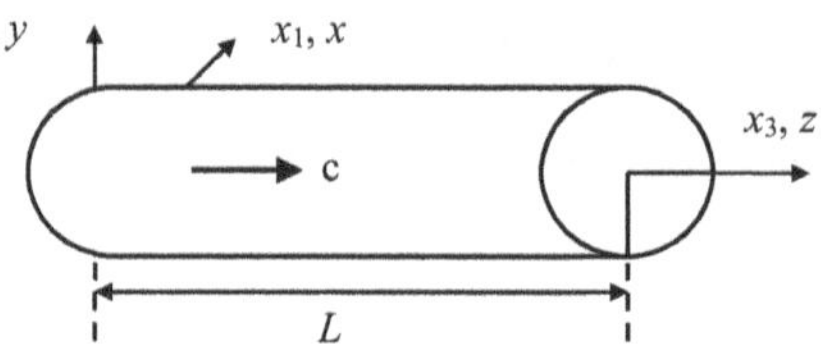

Fig. 4.1 A ZnO beam with its c-axis along x_3

$$J^p_{i,i} = -q\frac{\partial \Delta p}{\partial t},$$
$$J^n_{i,i} = q\frac{\partial \Delta n}{\partial t}, \tag{4.2}$$

$$S_{ij} = s^E_{ijkl}T_{kl} + d_{kij}E_k,$$
$$D_i = d_{ikl}T_{kl} + \varepsilon^T_{ik}E_k, \tag{4.3}$$

$$J^p_i = qp_0\mu^p_{ij}E_j - qD^p_{ij}(\Delta p)_{,j},$$
$$J^n_i = qn_0\mu^n_{ij}E_j + qD^n_{ij}(\Delta n)_{,j}, \tag{4.4}$$

$$S_{ij} = (u_{i,j} + u_{j,i})/2,$$
$$E_i = -\varphi_{,i}. \tag{4.5}$$

To develop a one-dimensional theory for the extension and bending in the *y*-*z* plane with shear deformation, we make the following approximations of the relevant mechanical displacements, electric potential, and carrier concentration perturbations [2]:

$$\begin{aligned}
u_2(\mathbf{x},t) &\cong v(x_3,t),\\
u_3(\mathbf{x},t) &\cong w(x_3,t) + x_2\psi(x_3,t),\\
\varphi(\mathbf{x},t) &\cong \phi^{(0)}(x_3,t) + x_2\phi^{(1)}(x_3,t),\\
\Delta p(\mathbf{x},t) &\cong p^{(0)}(x_3,t) + x_2p^{(1)}(x_3,t),\\
\Delta n(\mathbf{x},t) &\cong n^{(0)}(x_3,t) + x_2n^{(1)}(x_3,t),
\end{aligned} \tag{4.6}$$

where $w(x_3, t)$ is the extensional displacement, $v(x_3, t)$ the bending displacement, and $\psi(x_3, t)$ the shear deformation accompanying bending. Then, from Eq. (4.5), we calculate the relevant strains, electric fields, and carrier concentration perturbation gradients as

$$\begin{aligned}
&S_3 = S_{33} = u_{3,3} = w_{,3} + x_2\psi_{,3},\\
&S_4 = 2S_{23} = u_{2,3} + u_{3,2} = v_{,3} + \psi,\\
&E_2 = -\varphi_{,2} = -\phi^{(1)}, \quad E_3 = -\varphi_{,3} = -\phi^{(0)}_{,3} - x_2\phi^{(1)}_{,3},\\
&(\Delta p)_{,2} = p^{(1)}, \quad (\Delta p)_{,3} = p^{(0)}_{,3} + x_2p^{(1)}_{,3},\\
&(\Delta n)_{,2} = n^{(1)}, \quad (\Delta n)_{,3} = n^{(0)}_{,3} + x_2n^{(1)}_{,3}.
\end{aligned} \tag{4.7}$$

For bending in the y-z plane, the main stress components are T_3 and T_4. Therefore we introduce the following stress relaxation for thin beams:

$$T_1 = T_2 = T_5 = T_6 \cong 0. \tag{4.8}$$

From the constitutive relations in Eq. (4.3), for the relevant strain and electric displacement components, we have

$$\begin{aligned} S_3 &= s_{33}T_3 + d_{33}E_3, \\ S_4 &= s_{44}T_4 + d_{15}E_2, \\ D_2 &= d_{15}T_4 + \varepsilon_{11}E_2, \\ D_3 &= d_{33}T_3 + \varepsilon_{33}E_3. \end{aligned} \tag{4.9}$$

We invert Eq. $(4.9)_{1,2}$ for expressions of stresses in terms of strains and substitute the resulting expressions into Eq. $(4.9)_{3,4}$. This results in

$$\begin{aligned} T_3 &= T_{33} = \bar{c}_{33}S_3 - \bar{e}_{33}E_3 = \bar{c}_{33}\left(w_{,3} + x_2\psi_{,3}\right) + \bar{e}_{33}\left(\phi^{(0)}_{,3} + x_2\phi^{(1)}_{,3}\right), \\ T_4 &= T_{32} = \bar{c}_{44}S_4 - \bar{e}_{15}E_2 = \bar{c}_{44}(v_{,3} + \psi) + \bar{e}_{15}\phi^{(1)}, \\ D_2 &= \bar{e}_{15}S_4 + \bar{\varepsilon}_{11}E_2 = \bar{e}_{15}(v_{,3} + \psi) - \bar{\varepsilon}_{11}\phi^{(1)}, \\ D_3 &= \bar{e}_{33}S_3 + \bar{\varepsilon}_{33}E_3 = \bar{e}_{33}\left(w_{,3} + x_2\psi_{,3}\right) - \bar{\varepsilon}_{33}\left(\phi^{(0)}_{,3} + x_2\phi^{(1)}_{,3}\right), \end{aligned} \tag{4.10}$$

where Eq. (4.7) has been used and the one-dimensional effective material constants for thin beams are

$$\begin{aligned} &\bar{c}_{33} = 1/s_{33}, \quad \bar{c}_{44} = 1/s_{44}, \quad \bar{e}_{33} = d_{33}/s_{33}, \quad \bar{e}_{15} = d_{15}/s_{44}, \\ &\bar{\varepsilon}_{11} = \varepsilon_{11} - d_{15}^2/s_{44}, \quad \bar{\varepsilon}_{33} = \varepsilon_{33} - d_{33}^2/s_{33}. \end{aligned} \tag{4.11}$$

The relevant constitutive relations for the currents are the following ones from Eq. (4.4):

$$\begin{aligned} J_2^p &= qp_0\mu_{11}^pE_2 - qD_{11}^p(\Delta p)_{,2} = -qp_0\mu_{11}^p\phi^{(1)} - qD_{11}^pp^{(1)}. \\ J_3^p &= qp_0\mu_{33}^pE_3 - qD_{33}^p(\Delta p)_{,3} \\ &= -qp_0\mu_{33}^p\left(\phi^{(0)}_{,3} + x_2\phi^{(1)}_{,3}\right) - qD_{33}^p\left(p^{(0)}_{,3} + x_2p^{(1)}_{,3}\right), \\ J_2^n &= qn_0\mu_{11}^nE_2 + qD_{11}^n(\Delta n)_{,2} = -qn_0\mu_{11}^n\phi^{(1)} + qD_{11}^nn^{(1)}, \\ J_3^n &= qn_0\mu_{33}^nE_3 + qD_{33}^n(\Delta n)_{,3} \\ &= -qn_0\mu_{33}^n\left(\phi^{(0)}_{,3} + x_2\phi^{(1)}_{,3}\right) + qD_{33}^n\left(n^{(0)}_{,3} + x_2n^{(1)}_{,3}\right), \end{aligned} \tag{4.12}$$

where Eq. (4.7) has been used. Then the axial force N, the bending moment M, the transverse shear force Q, the zero- and first-order moments of the relevant electric displacement and current components can be expressed as

$$\begin{aligned}N &= \int_A T_3 dA = \bar{c}_{33}Aw_{,3} + \bar{e}_{33}A\phi^{(0)}_{,3},\\ M &= \int_A x_2T_3 dA = \bar{c}_{33}I\psi_{,3} + \bar{e}_{33}I\phi^{(1)}_{,3},\\ Q &= \int_A T_4 dA = \bar{c}_{44}A(v_{,3}+\psi) + \bar{e}_{15}A\phi^{(1)},\end{aligned} \tag{4.13}$$

$$\begin{aligned}D^{(0)}_2 &= \int_A D_2 dA = \bar{e}_{15}A(v_{,3}+\psi) - \bar{\varepsilon}_{11}A\phi^{(1)},\\ D^{(0)}_3 &= \int_A D_3 dA = \bar{e}_{33}Aw_{,3} - \bar{\varepsilon}_{33}A\phi^{(0)}_{,3},\\ D^{(1)}_3 &= \int_A x_2D_3 dA = \bar{e}_{33}I\psi_{,3} - \bar{\varepsilon}_{33}I\phi^{(1)}_{,3},\end{aligned} \tag{4.14}$$

$$\begin{aligned}J^{p(0)}_2 &= \int_A J^p_2 dA = -qp_0\mu^p_{11}A\phi^{(1)} - qD^p_{11}Ap^{(1)},\\ J^{p(0)}_3 &= \int_A J^p_3 dA = -qp_0\mu^p_{33}A\phi^{(0)}_{,3} - qD^p_{33}Ap^{(0)}_{,3},\\ J^{p(1)}_3 &= \int_A x_2J^p_3 dA = -qp_0\mu^p_{33}I\phi^{(1)}_{,3} - qD^p_{33}Ip^{(1)}_{,3},\end{aligned} \tag{4.15}$$

$$\begin{aligned}J^{n(0)}_2 &= \int_A J^n_2 dA = -qn_0\mu^n_{11}A\phi^{(1)} + qD^n_{11}An^{(1)},\\ J^{n(0)}_3 &= \int_A J^n_3 dA = -qn_0\mu^n_{33}A\phi^{(0)}_{,3} + qD^n_{33}An^{(0)}_{,3},\\ J^{n(1)}_3 &= \int_A x_2J^n_3 dA = -qn_0\mu^n_{33}I\phi^{(1)}_{,3} + qD^n_{33}In^{(1)}_{,3},\end{aligned} \tag{4.16}$$

where I and A are the moment of inertia and the area of the beam cross section. For a circular cross section, they are

$$I = \int_A x_2^2 dA = \frac{\pi a^4}{4}, \quad A = \pi a^2. \tag{4.17}$$

The one-dimensional equations of motion, the charge equation of electrostatics, and the conservation of charge for holes and electrons are obtained by integrating Eqs. (4.1) and (4.2) as well as their products with x_2 over the fiber cross section and using integration by parts or the two-dimensional divergence theorem over the cross section. The results are

$$\begin{aligned} N_{,3} &= \rho A\ddot{w}, \\ D_{3,3}^{(0)} &= qA\left(p^{(0)} - n^{(0)}\right), \\ J_{3,3}^{p(0)} &= -qA\dot{p}^{(0)}, \\ J_{3,3}^{n(0)} &= qA\dot{n}^{(0)}, \end{aligned} \tag{4.18}$$

$$\begin{aligned} &Q_{,3} = \rho A\ddot{v}, \\ &M_{,3} - Q = \rho I\ddot{\psi}, \\ &D_{3,3}^{(1)} - D_2^{(0)} = qI\left(p^{(1)} - n^{(1)}\right), \\ &J_{3,3}^{p(1)} - J_2^{p(0)} = -qI\dot{p}^{(1)}, \\ &J_{3,3}^{n(1)} - J_2^{n(0)} = qI\dot{n}^{(1)}. \end{aligned} \tag{4.19}$$

The substitution of Eqs. (4.13), (4.14), (4.15), and (4.16) into Eqs. (4.18) and (4.19) yields two sets of second-order linear ordinary differential equations. One is for extension involving w, $\phi^{(0)}$, $p^{(0)}$, and $n^{(0)}$ which is the same as the equations for extension used in multiple places in the previous chapter. The other is for bending with shear deformation involving v, ψ, $\phi^{(1)}$, $p^{(1)}$, and $n^{(1)}$ which is gathered below:

$$\begin{aligned} &\bar{c}_{44}A\left(v_{,33} + \psi_{,3}\right) + \bar{e}_{15}A\phi_{,3}^{(1)} = \rho A\ddot{v}, \\ &\bar{c}_{33}I\psi_{,33} + \bar{e}_{33}I\phi_{,33}^{(1)} - \bar{c}_{44}A(v_{,3} + \psi) - \bar{e}_{15}A\phi^{(1)} = \rho I\ddot{\psi}, \\ &\bar{e}_{33}I\psi_{,33} - \bar{\varepsilon}_{33}I\phi_{,33}^{(1)} - \bar{e}_{15}A(v_{,3} + \psi) + \bar{\varepsilon}_{11}A\phi^{(1)} = qI\left(p^{(1)} - n^{(1)}\right), \\ &-qp_0\mu_{33}^pI\phi_{,33}^{(1)} - qD_{33}^pIp_{,33}^{(1)} + qp_0\mu_{11}^pA\phi^{(1)} + qD_{11}^pAp^{(1)} = -qI\dot{p}^{(1)}, \\ &-qn_0\mu_{33}^nI\phi_{,33}^{(1)} + qD_{33}^nIn_{,33}^{(1)} + qn_0\mu_{11}^nA\phi^{(1)} - qD_{11}^nAn^{(1)} = qI\dot{n}^{(1)}. \end{aligned} \tag{4.20}$$

Shear deformation in a beam needs to be considered when the beam is not very long and thin, or when the wavelength is not much larger than the characteristic dimension of the cross section of the beam in a dynamic problem. The above equations are valid when the effect of shear deformation cannot be neglected in bending. Its accuracy can be improved with the use of a shear correction factor which is necessary when shear deformation is strong. Shear correction factors will be discussed in Chap. 5 on plates in a systematic manner.

4.2 Reduction to Bending Without Shear Deformation

For the beam in Fig. 4.1, we separate and collect the equations for bending with shear deformation from the previous section below, without the equations for extension. The mechanical displacements, electric potential, and carrier concentration perturbations are approximated by

$$\begin{aligned} u_2(\mathbf{x},t) &\cong v(x_3,t), \\ u_3(\mathbf{x},t) &\cong x_2\psi(x_3,t), \\ \varphi(\mathbf{x},t) &\cong x_2\phi^{(1)}(x_3,t), \\ \Delta p(\mathbf{x},t) &\cong x_2 p^{(1)}(x_3,t), \\ \Delta n(\mathbf{x},t) &\cong x_2 n^{(1)}(x_3,t). \end{aligned} \tag{4.21}$$

The field equations are

$$\begin{aligned} Q_{,3} &= \rho A\ddot{v}, \\ M_{,3} - Q &= \rho I\ddot{\psi}, \\ D^{(1)}_{3,3} - D^{(0)}_2 &= qI\left(p^{(1)} - n^{(1)}\right), \\ J^{p(1)}_{3,3} - J^{p(0)}_2 &= -qI\dot{p}^{(1)}, \\ J^{n(1)}_{3,3} - J^{n(0)}_2 &= qI\dot{n}^{(1)}. \end{aligned} \tag{4.22}$$

The constitutive relations are

$$\begin{aligned} M &= \int_A x_2 T_3 dA = \bar{c}_{33} I\psi_{,3} + \bar{e}_{33} I\phi^{(1)}_{,3}, \\ Q &= \int_A T_4 dA = \bar{c}_{44} A(v_{,3} + \psi) + \bar{e}_{15} A\phi^{(1)}, \end{aligned} \tag{4.23}$$

$$\begin{aligned} D^{(0)}_2 &= \int_A D_2 dA = \bar{e}_{15} A(v_{,3} + \psi) - \bar{\varepsilon}_{11} A\phi^{(1)}, \\ D^{(1)}_3 &= \int_A x_2 D_3 dA = \bar{e}_{33} I\psi_{,3} - \bar{\varepsilon}_{33} I\phi^{(1)}_{,3}, \end{aligned} \tag{4.24}$$

$$\begin{aligned} J^{p(0)}_2 &= \int_A J^p_2 dA = -qp_0\mu^p_{11} A\phi^{(1)} - qD^p_{11} Ap^{(1)}, \\ J^{p(1)}_3 &= \int_A x_2 J^p_3 dA = -qp_0\mu^p_{33} I\phi^{(1)}_{,3} - qD^p_{33} Ip^{(1)}_{,3}, \end{aligned} \tag{4.25}$$

$$\begin{aligned} J^{n(0)}_2 &= \int_A J^n_2 dA = -qn_0\mu^n_{11} A\phi^{(1)} + qD^n_{11} An^{(1)}, \\ J^{n(1)}_3 &= \int_A x_2 J^n_3 dA = -qn_0\mu^n_{33} I\phi^{(1)}_{,3} + qD^n_{33} In^{(1)}_{,3}. \end{aligned} \tag{4.26}$$

The above equations are for bending with shear deformation. For the special case of bending without shear deformation which is usually applicable to very thin beams, we set the beam shear strain $v_{,3} + \psi$ to zero, which implies that

$$\psi = -v_{,3}. \tag{4.27}$$

Then Eqs. $(4.23)_1$ and (4.24) reduce to

$$M = -\bar{c}_{33} I v_{,33} + \bar{e}_{33} I \phi^{(1)}_{,3}, \tag{4.28}$$

$$\begin{aligned} D_2^{(0)} &= -\bar{\varepsilon}_{11} A \phi^{(1)}, \\ D_3^{(1)} &= -\bar{e}_{33} I v_{,33} - \bar{\varepsilon}_{33} I \phi^{(1)}_{,3}. \end{aligned} \tag{4.29}$$

In bending without shear deformation, the moment of inertia I on the right-hand side of Eq. $(4.22)_2$ can be neglected. Then Eq. $(4.22)_2$ implies that

$$Q = M_{,3} = -\bar{c}_{33} I v_{,333} + \bar{e}_{33} I \phi^{(1)}_{,33}, \tag{4.30}$$

where Eq. (4.28) has been used. Effectively, Eq. (4.30) serves as the constitutive relation for Q, and Eq. $(4.23)_2$ is abandoned. When Eq. (4.30) is substituted into Eq. $(4.22)_1$, we obtain the equation for bending without shear deformation as

$$-\bar{c}_{33} I v_{,3333} + \bar{e}_{33} I \phi^{(1)}_{,333} = \rho A \ddot{v}. \tag{4.31}$$

Equation (4.31) is accompanied by Eq. $(4.22)_{3\text{–}5}$ to produce four equations needed for v, $\phi^{(1)}$, $p^{(1)}$, and $n^{(1)}$.

4.3 Static Bending of a Cantilever

Consider the ZnO beam shown in Fig. 4.2. The left end is fixed. The right end is under the action of a transverse shear force f_y [2].

We consider bending with shear deformation governed by Eqs. (4.22)–(4.26) and (4.20). We limit ourselves to the case of an n-type semiconductor so that $p^{(1)} \cong 0$. The four remaining fields are v, ψ, $\phi^{(1)}$, and $n^{(1)}$. For static bending, the relevant equations reduce to

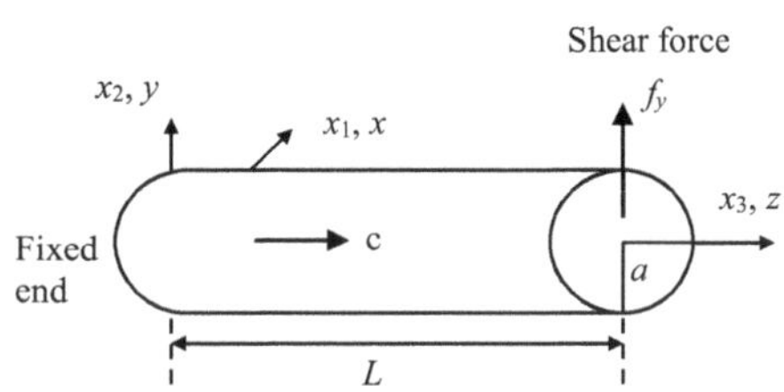

Fig. 4.2 A ZnO beam under a static shear force f_y

$$
\begin{aligned}
&Q_{,3} = 0, \\
&M_{,3} - Q = 0, \\
&D_{3,3}^{(1)} - D_2^{(0)} = qI\left(p^{(1)} - n^{(1)}\right), \\
&J_{3,3}^{n(1)} - J_2^{n(0)} = 0,
\end{aligned} \tag{4.32}
$$

$$
\begin{aligned}
&Q = \bar{c}_{44}A(v_{,3} + \psi) + \bar{e}_{15}A\phi^{(1)}, \\
&M = \bar{c}_{33}I\psi_{,3} + \bar{e}_{33}I\phi_{,3}^{(1)},
\end{aligned} \tag{4.33}
$$

$$
\begin{aligned}
&D_2^{(0)} = \bar{e}_{15}A(v_{,3} + \psi) - \bar{\varepsilon}_{11}A\phi^{(1)}, \\
&D_3^{(1)} = \bar{e}_{33}I\psi_{,3} - \bar{\varepsilon}_{33}I\phi_{,3}^{(1)},
\end{aligned} \tag{4.34}
$$

$$
\begin{aligned}
&J_2^{n(0)} = -qn_0\mu_{11}^n A\phi^{(1)} + qD_{11}^n A n^{(1)}, \\
&J_3^{n(1)} = -qn_0\mu_{33}^n I\phi_{,3}^{(1)} + qD_{33}^n I n_{,3}^{(1)},
\end{aligned} \tag{4.35}
$$

$$
\begin{aligned}
&\bar{c}_{44}A\left(v_{,33} + \psi_{,3}\right) + \bar{e}_{15}A\phi_{,3}^{(1)} = 0, \\
&\bar{c}_{33}I\psi_{,33} + \bar{e}_{33}I\phi_{,33}^{(1)} - \bar{c}_{44}A(v_{,3} + \psi) - \bar{e}_{15}A\phi^{(1)} = 0, \\
&\bar{e}_{33}I\psi_{,33} - \bar{\varepsilon}_{33}I\phi_{,33}^{(1)} - \bar{e}_{15}A(v_{,3} + \psi) + \bar{\varepsilon}_{11}A\phi^{(1)} = qI\left(-n^{(1)}\right), \\
&-qn_0\mu_{33}^n I\phi_{,33}^{(1)} + qD_{33}^n I n_{,33}^{(1)} + qn_0\mu_{11}^n A\phi^{(1)} - qD_{11}^n A n^{(1)} = 0.
\end{aligned} \tag{4.36}
$$

The boundary conditions are

$$
v(0) = 0, \quad \psi(0) = 0, \quad M(L) = 0, \quad Q(L) = f_y, \tag{4.37}
$$

$$
D_3^{(1)}(0) = 0, \quad J_3^{n(1)}(0) = 0, \quad D_3^{(1)}(L) = 0, \quad J_3^{n(1)}(L) = 0, \tag{4.38}
$$

where we have assumed an electrically isolated beam. For bending without extension, the carrier concentration perturbation is simply $\Delta n = x_2 n^{(1)}$, an odd function of x_2 that satisfies the charge neutrality condition automatically.

With some algebra, it can be found that the general solution of Eq. (4.36) is

$$
\begin{aligned}
\phi^{(1)} =& C_1 \sinh\lambda_1 x_3 + C_2 \cosh\lambda_1 x_3 + C_3 \sinh\lambda_3 x_3 + C_4 \cosh\lambda_3 x_3 \\
&- \frac{D_{11}^n A}{D_{33}^n \varepsilon_{33} I}\left(\frac{\bar{e}_{33}\bar{c}_{44} - \bar{e}_{15}\bar{c}_{33}}{\bar{c}_{44}}\right)\frac{C_5}{a_2},
\end{aligned} \tag{4.39}
$$

$$
\begin{aligned}
\psi =& -\frac{\bar{e}_{33}}{\bar{c}_{33}}(C_1 \sinh\lambda_1 x_3 + C_2 \cosh\lambda_1 x_3 + C_3 \sinh\lambda_3 x_3 + C_4 \cosh\lambda_3 x_3) \\
&+ \frac{C_5}{2}x_3^2 + C_6 x_3 + C_7 + \frac{D_{11}^n A\bar{e}_{33}}{D_{33}^n \varepsilon_{33} I\bar{c}_{33}}\left(\frac{\bar{e}_{33}\bar{c}_{44} - \bar{e}_{15}\bar{c}_{33}}{\bar{c}_{44}}\right)\frac{C_5}{a_2},
\end{aligned} \tag{4.40}
$$

$$
\begin{aligned}
n^{(1)} = \frac{1}{qI}&[(\varepsilon_{33}I\lambda_1^2 - \varepsilon_{11}A)(C_1 \sinh \lambda_1 x_3 + C_2 \cosh \lambda_1 x_3) \\
&+ (\varepsilon_{33}I\lambda_3^2 - \varepsilon_{11}A)(C_3 \sinh \lambda_3 x_3 + C_4 \cosh \lambda_3 x_3) \\
&- \frac{n_0 \mu_{11}^n I(\bar{e}_{33}\bar{c}_{44} - \bar{e}_{15}\bar{c}_{33})}{(q n_0 \mu_{11}^n I + D_{11}^n \varepsilon_{11} A)\bar{c}_{44}} C_5],
\end{aligned}
\tag{4.41}
$$

$$
\begin{aligned}
v = &\left(\frac{\bar{e}_{33}}{\bar{c}_{33}} - \frac{\bar{e}_{15}}{\bar{c}_{44}}\right)\left[\frac{1}{\lambda_1}(C_1 \cosh \lambda_1 x_3 + C_2 \sinh \lambda_1 x_3)\right. \\
&\left. + \frac{1}{\lambda_3}(C_3 \cosh \lambda_3 x_3 + C_4 \sinh \lambda_3 x_3)\right] - \frac{C_5}{6}x_3^3 - \frac{C_6}{2}x_3{}^2 \\
&+ \left[\left(\frac{\bar{c}_{33}I}{\bar{c}_{44}A} - \frac{D_{11}^n A(\bar{e}_{33}\bar{c}_{44} - \bar{e}_{15}\bar{c}_{33})^2}{D_{33}^n \varepsilon_{33} I \bar{c}_{33} \bar{c}_{44}^2 a_2}\right)C_5 - C_7\right]x_3 + C_8,
\end{aligned}
\tag{4.42}
$$

where C_1 through C_8 are eight arbitrary constants. λ_1 through λ_4 are the four roots of the following equation:

$$
\lambda^4 - a_1\lambda^2 + a_2 = 0, \tag{4.43}
$$

and

$$
a_1 = \frac{q\mu_{33}^n n_0}{D_{33}^n \varepsilon_{33}} + \frac{(D_{33}^n \varepsilon_{11} + D_{11}^n \varepsilon_{33})A}{D_{33}^n \varepsilon_{33} I}, \quad a_2 = \frac{q\mu_{11}^n A n_0}{D_{33}^n \varepsilon_{33} I} + \frac{D_{11}^n \varepsilon_{11} A^2}{D_{33}^n \varepsilon_{33} I^2}. \tag{4.44}
$$

Substituting Eqs. (4.39), (4.40), (4.41), and (4.42) into the boundary conditions in Eqs. (4.37) and (4.38), we obtain eight linear algebraic equations for C_1 through C_8. These equations are solved on a computer.

As an example, consider a beam whose geometric parameters are that $L = 600$ nm and $a = 25$ nm. $n_0 = N_D^+ = 10^{23}$ m^{-3}. $D_{11}^n = D_{33}^n$ is used in the calculation because of insufficient material data [2]. For these parameters, λ_1 through λ_4 are all real and so are Eqs. (4.39), (4.40), (4.41), and (4.42). In addition, $\lambda_1 = -\lambda_2$ and $\lambda_3 = -\lambda_4$. Numerical results show that when $f_y < 0.2$ nN, Δn is at least an order of magnitude smaller than n_0, and the linearized theory is valid.

$\Delta n = x_2 n^{(1)}$ shows that the behavior of the electron concentration perturbation is determined by $n^{(1)}$. In Fig. 4.3, $n^{(1)}$ is plotted for different values of $f = f_y$. (a) shows that $n^{(1)}$ is nearly constant except near the left end where it is large and varies rapidly. For the relevant applications, its behavior away from the fixed end is of interest. To show the behavior of $n^{(1)}$ away from the fixed end more clearly, it is plotted in (b) without a small region near the fixed end. It can be seen that a larger end force corresponds to a larger $n^{(1)}$ as expected. From Eqs. (4.33) and (4.34), it can be seen that this electron distribution is caused by the shear force Q or the related shear stress T_4 through $e_{24} = e_{15}$ and the bending moment M as well as $D_3^{(1)}$ through $\bar{e}_{33}$. The behavior of $\phi^{(1)}$ is found to be similar to that of $n^{(1)}$ and therefore is not shown here.

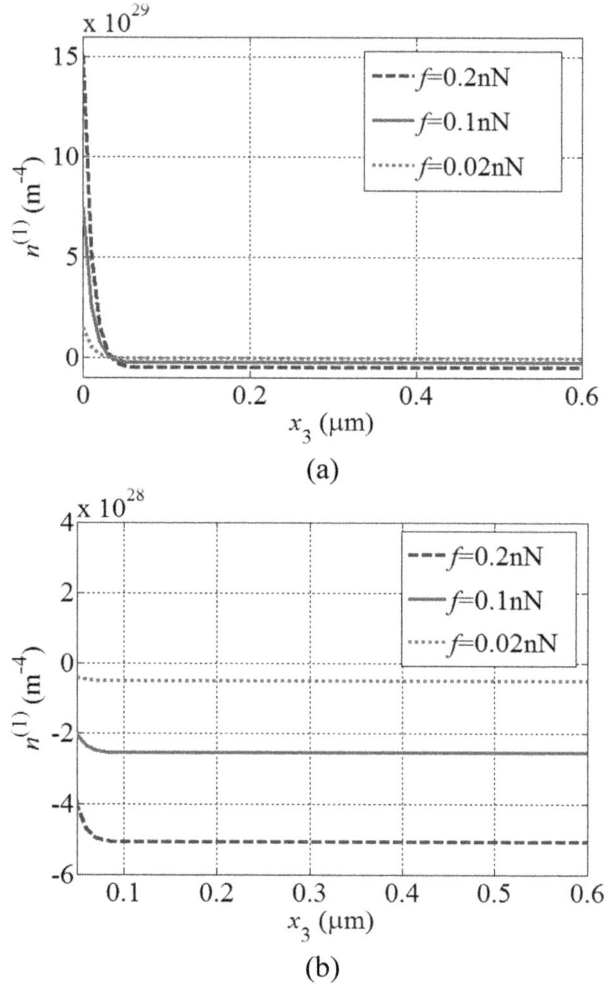

Fig. 4.3 Effects of the end force $f = f_y$ on electron concentration perturbation $n^{(1)}$. $n_0 = 10^{23}/\mathrm{m}^3$. (**a**) $0 \le x_3 \le 600$ nm. (**b**) $50 \le x_3 \le 600$ nm

4.4 Time-Harmonic Bending

Consider the ZnO beam shown in Fig. 4.4. The shear force $f_y(t)$ is assumed to be time-harmonic in this section [3].

We consider bending with shear deformation. We limit ourselves to the case of an n-type semiconductor so that $p^{(1)}$ vanishes. The four remaining fields are v, ψ, $\phi^{(1)}$, and $n^{(1)}$. The relevant equations from Eqs. (4.22)–(4.26) and (4.20) are

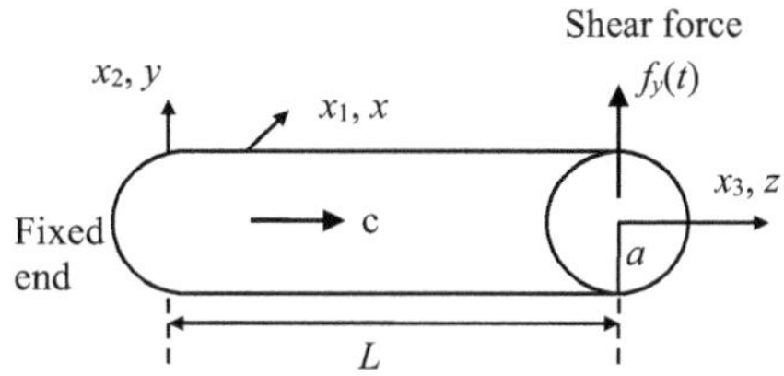

Fig. 4.4 A ZnO beam under a time-harmonic shear force

$$\begin{aligned} &Q_{,3} = \rho A\ddot{v}, \\ &M_{,3} - Q = \rho I\ddot{\psi}, \\ &D^{(1)}_{3,3} - D^{(0)}_{2} = qI\left(-n^{(1)}\right), \\ &J^{n(1)}_{3,3} - J^{n(0)}_{2} = qI\dot{n}^{(1)}, \end{aligned} \tag{4.45}$$

$$\begin{aligned} Q &= \bar{c}_{44}A(v_{,3} + \psi) + \bar{e}_{15}A\phi^{(1)}, \\ M &= \bar{c}_{33}I\psi_{,3} + \bar{e}_{33}I\phi^{(1)}_{,3}, \end{aligned} \tag{4.46}$$

$$\begin{aligned} D^{(0)}_{2} &= \bar{e}_{15}A(v_{,3} + \psi) - \bar{\varepsilon}_{11}A\phi^{(1)}, \\ D^{(1)}_{3} &= \bar{e}_{33}I\psi_{,3} - \bar{\varepsilon}_{33}I\phi^{(1)}_{,3}, \end{aligned} \tag{4.47}$$

$$\begin{aligned} J^{n(0)}_{2} &= -qn_0\mu^n_{11}A\phi^{(1)} + qD^n_{11}An^{(1)}, \\ J^{n(1)}_{3} &= -qn_0\mu^n_{33}I\phi^{(1)}_{,3} + qD^n_{33}In^{(1)}_{,3}, \end{aligned} \tag{4.48}$$

$$\begin{aligned} &\bar{c}_{44}A\left(v_{,33} + \psi_{,3}\right) + \bar{e}_{15}A\phi^{(1)}_{,3} = \rho A\ddot{v}, \\ &\bar{c}_{33}I\psi_{,33} + \bar{e}_{33}I\phi^{(1)}_{,33} - \bar{c}_{44}A(v_{,3} + \psi) - \bar{e}_{15}A\phi^{(1)} = \rho I\ddot{\psi}, \\ &\bar{e}_{33}I\psi_{,33} - \bar{\varepsilon}_{33}I\phi^{(1)}_{,33} - \bar{e}_{15}A(v_{,3} + \psi) + \bar{\varepsilon}_{11}A\phi^{(1)} = qI\left(-n^{(1)}\right), \\ &-qn_0\mu^n_{33}I\phi^{(1)}_{,33} + qD^n_{33}In^{(1)}_{,33} + qn_0\mu^n_{11}A\phi^{(1)} - qD^n_{11}An^{(1)} = qI\dot{n}^{(1)}. \end{aligned} \tag{4.49}$$

The boundary conditions are

$$v(0) = 0, \quad \psi(0) = 0, \quad M(L) = 0, \quad Q(L) = f_y(t) = F\exp(i\omega t), \tag{4.50}$$

$$D^{(1)}_3(0) = 0, \quad J^{n(1)}_3(0) = 0, \quad D^{(1)}_3(L) = 0, \quad J^{n(1)}_3(L) = 0, \tag{4.51}$$

where F and ω are real constants and i is the imaginary unit. We use the usual complex notation with

$$\left\{v, \psi, \phi^{(1)}, n^{(1)}\right\} = \mathrm{Re}\{\{\mathrm{V}, \Psi, \Phi, \mathrm{N}\}\exp(i\omega t)\}. \tag{4.52}$$

Then Eq. (4.49) reduces to the following four ordinary differential equations for V, Ψ, Φ, and N:

$$\begin{aligned}
&\bar{c}_{44}A(\mathrm{V}_{,33}+\Psi_{,3})+\bar{e}_{15}A\Phi_{,3}+\rho\omega^2 A\mathrm{V}=0,\\
&\bar{c}_{33}I\Psi_{,33}+\bar{e}_{33}I\Phi_{,33}-\bar{c}_{44}A(\mathrm{V}_{,3}+\Psi)-\bar{e}_{15}A\Phi+\rho\omega^2 I\Psi=0,\\
&\bar{e}_{33}I\Psi_{,33}-\bar{\varepsilon}_{33}I\Phi_{,33}-\bar{e}_{15}A(\mathrm{V}_{,3}+\Psi)+\bar{\varepsilon}_{11}A\Phi+qI\mathrm{N}=0,\\
&-qn_0\mu_{33}^n I\Phi_{,33}+qD_{33}^n I\mathrm{N}_{,33}+qn_0\mu_{11}^n A\Phi-qD_{11}^n A\mathrm{N}-i\omega qI\mathrm{N}=0.
\end{aligned} \tag{4.53}$$

We look for a solution to Eq. (4.53) in the following form:

$$\{\mathrm{V},\Psi,\Phi,\mathrm{N}\}=\{C_1,C_2,C_3,C_4\}\exp(\lambda x_3). \tag{4.54}$$

The substitution of Eq. (4.54) into Eq. (4.53) results in four linear homogeneous algebraic equations for C_1–C_4. For nontrivial solutions, the determinant of the coefficient matrix of the equations has to vanish, which leads to a polynomial equation of degree eight for λ. Let the eight roots of this equation be $\lambda^{(m)}$, and the corresponding nontrivial solution of C_1–C_4 be $C_1^{(m)}-C_4^{(m)}$. Since Eq. (4.53) is homogeneous, only the ratios among $C_1^{(m)}-C_4^{(m)}$ can be determined. Then the general solution to Eq. (4.53) can be written as

$$\begin{Bmatrix}\mathrm{V}\\ \Psi\\ \Phi\\ \mathrm{N}\end{Bmatrix}=\sum_{m=1}^{8}G^{(m)}\begin{Bmatrix}C_1^{(m)}\\ C_2^{(m)}\\ C_3^{(m)}\\ C_4^{(m)}\end{Bmatrix}\exp\left(\lambda^{(m)}x_3\right), \tag{4.55}$$

where $G^{(m)}$ are undetermined constants. Substituting Eq. (4.55) into Eqs. (4.50) and (4.51), we obtain eight linear algebraic equations for $G^{(m)}$, which are solved on a computer using MATLAB.

As an example, consider a circular ZnO beam with $L = 600$ nm and $a = 25$ nm. $n_0 = N_D^+ = 10^{23}\ \mathrm{m}^{-3}$. Material damping is included by allowing the elastic constants to assume complex values which can describe viscous damping in the material. In our numerical calculation, the real elastic constant c_{pq} are replaced by $c_{pq}(1 + iQ^{-1})$, where Q is the quality factor of the material, a large, real and positive number. We fix Q to be 10^2 in our calculation. In the static bending analysis in the previous section, end forces greater than or equal to 0.02 nN were used. In this section, since we are interested in the behavior of the beam near resonances where a small driving force can produce significant motions, in our calculation below we reduce the end force to $F = 0.002$ nN so that even at resonances all of the fields are small, and the linear theory used for small deformation is still valid.

We plot the modulus (absolute value) of the complex deflection at the right end versus the driving frequency in Fig. 4.5. In the frequency range shown, there are three resonances at $\omega_{(1)} = 6.1570 \times 10^8$rad/s, $\omega_{(2)} = 3.7651 \times 10^9$rad/s, and $\omega_{(3)} = 1.0171 \times 10^{10}$ rad/s where the right end displacement is large. The first peak is relatively more pronounced. The second and the third peaks are barely visible

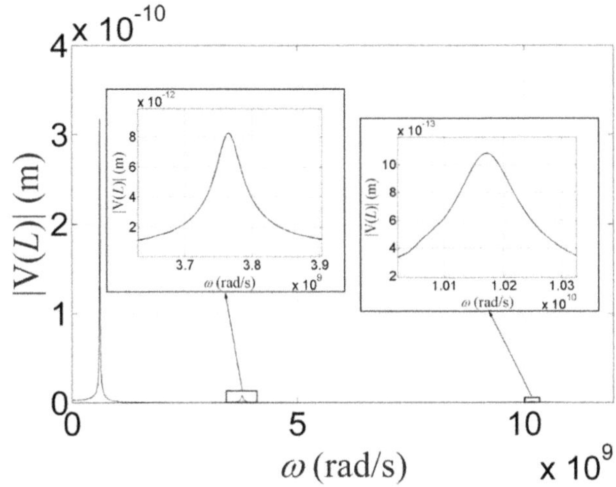

Fig. 4.5 Beam end displacement versus driving frequency showing three resonances

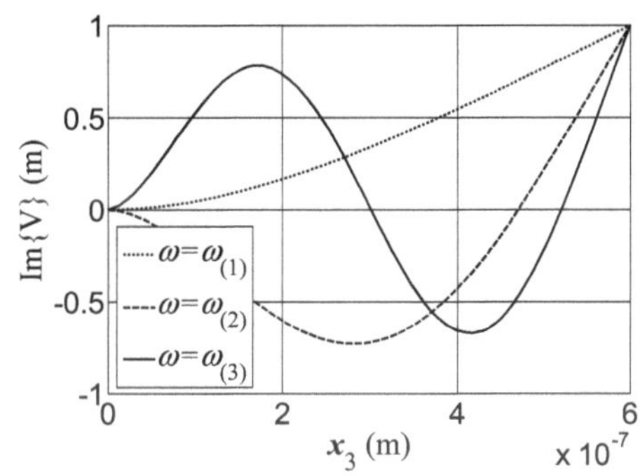

Fig. 4.6 Normalized imaginary parts of the complex deflection $V(x_3)$ along the beam at resonances

and are magnified locally in the figure. Below is the first resonance frequency where there is a relatively small but finite response, which is the low-frequency limit.

For the deflection curve at the three resonances, it is found that the imaginary parts are much larger than the real parts. Therefore we focus on the imaginary parts and normalize them in Fig. 4.6. The behavior is familiar for a cantilever at resonances. In addition to the fixed left end, the three displacement distributions are with zero, one, and two points with zero deflection (nodal points) as expected.

We are interested in the distribution of the carrier concentration perturbation described by $n^{(1)}x_2$ or Nx_2. Numerical results show that its imaginary part is much larger than the real part. Therefore, only its imaginary part is shown in Fig. 4.7. The carrier concentration perturbation becomes very large at the left end which may not be true because it is well-known that one-dimensional theories for the bending of a beam cannot predict behaviors at a fixed end accurately in general. For the most part of the beam away from its ends, the one-dimensional theory can produce reliable

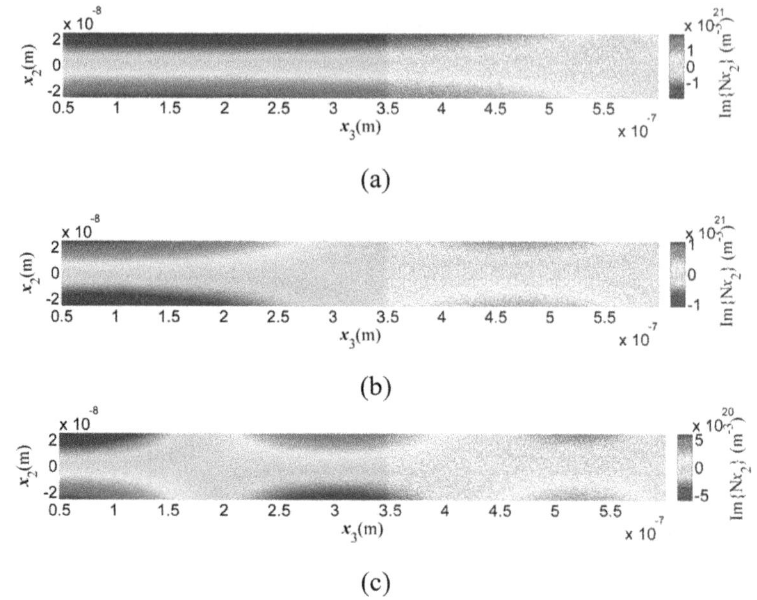

Fig. 4.7 Imaginary parts of the carrier concentration distribution perturbation Nx_2 at resonances. (**a**) $\omega = \omega_{(1)}$. (**b**) $\omega = \omega_{(2)}$. (**c**) $\omega = \omega_{(3)}$

results. Since the fields are large and may be inaccurate near the left end at $x_3 = 0$, only the results for $x_3 > 0.5 \times 10^{-7}$ m are shown. It is seen that at the first resonance the carrier concentration is mainly near the top and bottom surfaces and is essentially uniform along the beam except near the right end. At the second and the third resonances, the fields have nodal points (zeros) along x_3 and change their signs across the nodal points. The behavior of the electric potential is similar [3].

4.5 Transient Bending

Consider an n-type ZnO beam with a circular cross section as shown in Fig. 4.8. The left end is fixed. The right end is under the action of a transverse shear force f_y suddenly applied when the beam is at rest at $t = 0$ [4].

Consider bending without shear deformation. The relevant equations from Sect. 4.2 are

$$\begin{aligned} &u_2(\mathbf{x}, t) \cong v(x_3, t), \quad u_3(\mathbf{x}, t) \cong -x_2 v_{,3}(x_3, t), \\ &\varphi(\mathbf{x}, t) \cong x_2 \phi^{(1)}(x_3, t), \quad \Delta n(\mathbf{x}, t) \cong x_2 n^{(1)}(x_3, t), \end{aligned} \tag{4.56}$$

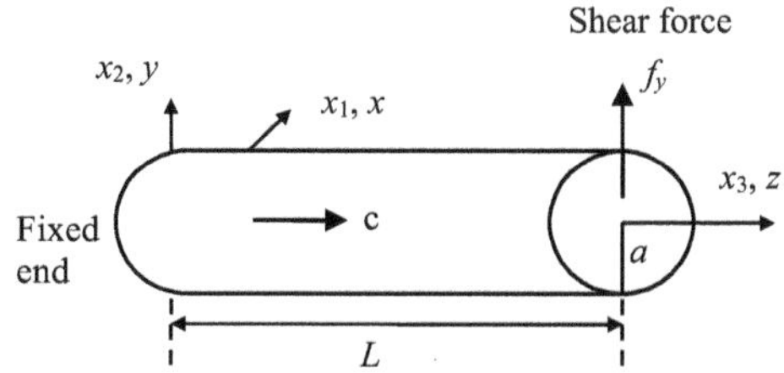

Fig. 4.8 A ZnO beam under a suddenly applied shear force

$$Q_{,3} = \rho A\ddot{v}, \tag{4.57}$$

$$\begin{aligned} D_{3,3}^{(1)} - D_2^{(0)} &= qI\left(-n^{(1)}\right), \\ J_{3,3}^{n(1)} - J_2^{n(0)} &= qI\dot{n}^{(1)}, \end{aligned} \tag{4.58}$$

$$M = -\overline{c}_{33}Iv_{,33} + \overline{e}_{33}I\phi_{,3}^{(1)}, \tag{4.59}$$

$$Q = M_{,3} = -\overline{c}_{33}Iv_{,333} + \overline{e}_{33}I\phi_{,33}^{(1)}, \tag{4.60}$$

$$\begin{aligned} D_2^{(0)} &= -\overline{\varepsilon}_{11}A\phi^{(1)}, \\ D_3^{(1)} &= -\overline{e}_{33}Iv_{,33} - \overline{\varepsilon}_{33}I\phi_{,3}^{(1)}, \end{aligned} \tag{4.61}$$

$$\begin{aligned} J_2^{n(0)} &= -qn_0\mu_{11}^n A\phi^{(1)} + qD_{11}^n An^{(1)}, \\ J_3^{n(1)} &= -qn_0\mu_{33}^n I\phi_{,3}^{(1)} + qD_{33}^n In_{,3}^{(1)}. \end{aligned} \tag{4.62}$$

Substituting Eqs. (4.60), (4.61), and (4.62) into Eqs. (4.57) and (4.58), we obtain the following equations for v, $\phi^{(1)}$, and $n^{(1)}$:

$$-\overline{c}_{33}Iv_{,3333} + \overline{e}_{33}I\phi_{,333}^{(1)} = \rho A\ddot{v}, \tag{4.63}$$

$$-\overline{e}_{33}Iv_{,333} - \overline{\varepsilon}_{33}I\phi_{,33}^{(1)} + \overline{\varepsilon}_{11}A\phi^{(1)} = qI\left(-n^{(1)}\right), \tag{4.64}$$

$$-n_0\mu_{33}^n I\phi_{,33}^{(1)} + D_{33}^n In_{,33}^{(1)} + n_0\mu_{11}^n A\phi^{(1)} - D_{11}^n An^{(1)} = I\dot{n}^{(1)}. \tag{4.65}$$

We need to solve Eqs. (4.63), (4.64), and (4.65) with the following boundary and initial conditions:

$$\begin{aligned} &v = 0, \quad v_{,3} = 0, \quad D_3^{(1)} = 0, \quad J_3^{n(1)} = 0, \quad x_3 = 0, \\ &M = 0, \quad Q = f_y, \quad D_3^{(1)} = 0, \quad J_3^{n(1)} = 0, \quad x_3 = L, \end{aligned} \tag{4.66}$$

$$v = 0, \quad \dot{v} = 0, \quad \phi^{(1)} = 0, \quad n^{(1)} = 0, \quad t = 0. \tag{4.67}$$

An exact theoretical analysis is very challenging mathematically if not impossible. In the following, we make a few approximations to break Eqs. (4.63), (4.64), and (4.65) into three subproblems that are one-way coupled and solve them sequentially.

We begin the approximations by neglecting the piezoelectric coupling in the bending equation in Eq. (4.63) and obtain the following initial-boundary-value problem for the bending of an elastic beam:

$$\begin{aligned} &\text{-}\bar{c}_{33} I v_{,3333} = \rho A \ddot{v}, \\ &v = 0, \quad v_{,3} = 0, \quad x_3 = 0, \\ &M = 0, \quad Q = f_y, \quad x_3 = L, \\ &v = 0, \quad \dot{v} = 0, \quad t = 0. \end{aligned} \tag{4.68}$$

To make the shear force boundary condition in Eq. (4.68) homogeneous, we make a change of the unknown variable from v to $\widehat{v}$ through

$$v(x_3, t) = \widehat{v}(x_3, t) - \frac{f_y}{6\bar{c}_{33} I} x_3^3 + \frac{f_y L}{2\bar{c}_{33} I} x_3^2. \tag{4.69}$$

Then Eq. (4.68) becomes

$$\begin{aligned} &-\bar{c}_{33} I \widehat{v}_{,3333} = \rho A \ddot{\widehat{v}}, \\ &\widehat{v} = 0, \quad \widehat{v}_{,3} = 0, \quad x_3 = 0, \\ &M = 0, \quad Q = 0, \quad x_3 = L, \\ &\widehat{v} = \frac{f_y}{6\bar{c}_{33} I} x_3^3 - \frac{f_y L}{2\bar{c}_{33} I} x_3^2, \quad \dot{\widehat{v}} = 0, \quad t = 0. \end{aligned} \tag{4.70}$$

Equation (4.70) can be solved using the method of separation of variables without any challenge. We skip the procedure and summarize the solution below:

$$v(x_3, t) = \sum_{m=1}^{\infty} A^{(m)} Y^{(m)}(x_3) \cos\left(\omega^{(m)} t\right) - \frac{f_y}{6\bar{c}_{33} I} x_3^3 + \frac{f_y L}{2\bar{c}_{33} I} x_3^2, \tag{4.71}$$

where

$$\begin{aligned} Y^{(m)}(x_3) &= \cos\beta^{(m)} x_3 - \cosh\beta^{(m)} x_3 \\ &\quad + \xi^{(m)}\left(\sin\beta^{(m)} x_3 - \sinh\beta^{(m)} x_3\right), \\ \xi^{(m)} &= -\frac{\cos\beta^{(m)} L + \cosh\beta^{(m)} L}{\sin\beta^{(m)} L + \sinh\beta^{(m)} L}, \end{aligned} \tag{4.72}$$

are the free vibration modes. $\beta^{(m)}$ is the roots of

$$\cos \beta L \cdot \cosh \beta L + 1 = 0, \tag{4.73}$$

and

$$\left(\omega^{(m)}\right)^2 = \frac{\bar{c}_{33} I}{\rho \mathrm{A}} \left(\beta^{(m)}\right)^4. \tag{4.74}$$

In Eq. (4.71), $A^{(m)}$ is determined from the initial conditions.

With $v(x_3, t)$ known, we make another approximation by neglecting $n^{(1)}$ in Eq. (4.64) and obtain the following boundary-value problem for $\phi^{(1)}$:

$$\begin{aligned} &\phi^{(1)}_{,33} - \lambda^2 \phi^{(1)} = -\frac{\bar{e}_{33}}{\bar{\varepsilon}_{33}} v_{,333}, \\ &D_3^{(1)} = 0, \quad x_3 = 0, \ L, \end{aligned} \tag{4.75}$$

where

$$\lambda^2 = \frac{\bar{\varepsilon}_{11} A}{\bar{\varepsilon}_{33} I}. \tag{4.76}$$

The general solution to Eq. $(4.75)_1$ can be formally written as

$$\phi^{(1)} = C_1(t) \exp(\lambda x_3) + C_2(t) \exp(-\lambda x_3) + \phi^p(x_3, t), \tag{4.77}$$

where $C_1(t)$ and $C_2(t)$ are determined from the boundary conditions in Eq. $(4.75)_2$. $\phi^p(x_3, t)$ is a particular solution to Eq. $(4.75)_1$ in the form of a linear combination of $\cos\beta^{(m)} x_3$, $\sin\beta^{(m)} x_3$, $\cosh\beta^{(m)} x_3$, and $\sinh\beta^{(m)} x_3$. The coefficients of the linear combination depend on $\cos\omega^{(m)} t$. The expressions of $C_1(t)$, $C_2(t)$, and $\phi^p(x_3, t)$ are lengthy and are not presented here.

With $\phi^{(1)}(x, t)$ known, from Eqs. (4.65), (4.66), and (4.67), we have the following initial-boundary-value problem for $n^{(1)}$:

$$\begin{aligned} &-n_0 \mu^n_{33} I \phi^{(1)}_{,33} + D^n_{33} I n^{(1)}_{,33} + n_0 \mu^n_{11} A \phi^{(1)} - D^n_{11} A n^{(1)} = I \dot{n}^{(1)}, \\ &J_3^{n(1)} = 0, \quad x_3 = 0, \ L, \\ &n^{(1)} = 0, \quad t = 0. \end{aligned} \tag{4.78}$$

To make the boundary conditions in Eq. (4.78) homogeneous, we let

$$n^{(1)}(x, t) = \widehat{n}(x, t) + \frac{n_0 \mu^n_{33}}{D^n_{33}} \phi^{(1)}(x, t). \tag{4.79}$$

The mathematical problem for $\widehat{n}$ is

$$
\begin{aligned}
&D_{33}^n I\hat{n}_{,33} - D_{11}^n A\hat{n} - I\dot{\hat{n}} = \frac{n_0\mu_{33}^n}{D_{33}^n} I\dot{\phi}^{(1)} + \left(D_{11}^n \frac{\mu_{33}^n}{D_{33}^n} - \mu_{11}^n\right) A n_0 \phi^{(1)}, \\
&\hat{n}_{,3} = 0, \quad x_3 = 0, L, \\
&\hat{n} = 0, \quad t = 0.
\end{aligned}
\tag{4.80}
$$

Equation (4.80) is a standard mathematical problem. Its solution can be obtained by separation of variables without any challenge, and is not presented here.

As an example, consider a circular ZnO beam with $L = 600$ nm and $a = 25$ nm. The suddenly applied end shear force $f_y = 0.2$ nN. $n_0 = N_D^+ = 10^{23}$ m^{-3}. Right after the load is applied, the beam rises from the horizontal reference state, gradually reaches its largest deflection, then falls back. The initial rising part of the motion is shown in Fig. 4.9 for a few time instants. Five terms of the series are used in Fig. 4.9. This is very accurate for predicting the deflection curve. However, for the potential and carrier concentration to be calculated next, it is less accurate [4]. We note that there is a fundamental difference between the extensional waves in Sect. 3.12 and the bending waves here. When a load is suddenly applied at the right end of the beam, it generates component waves or modes with different frequencies and wavelengths. Extensional waves are nondispersive. All waves propagate at the same finite speed. Therefore the left end feels the initial disturbance at the right end after a finite time interval. On the contrary, bending waves are dispersive. Long waves propagate slowly, but short waves propagate quickly. Therefore, for bending, the left end of the beam quickly feels the initial disturbance at the right end, essentially taking no time.

We are particularly interested in the development of the electron distribution under $\phi^{(1)}$. In Fig. 4.10, we show $\Delta n = x_2 n^{(1)}(x_3, t)$ at the four time instants during the initial rising part of motion in the region of interest, i.e., away from the left end where there are field concentrations. Figure 4.10 is in fact from a COMSOL numerical solution of Eq. (4.80) [4]. The analytical solution of Eq. (4.80) looks

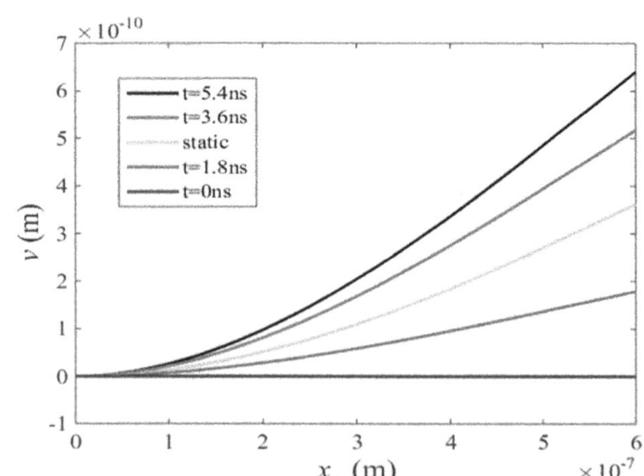

Fig. 4.9 Deflection curve at different time instants. $m = 1$–5

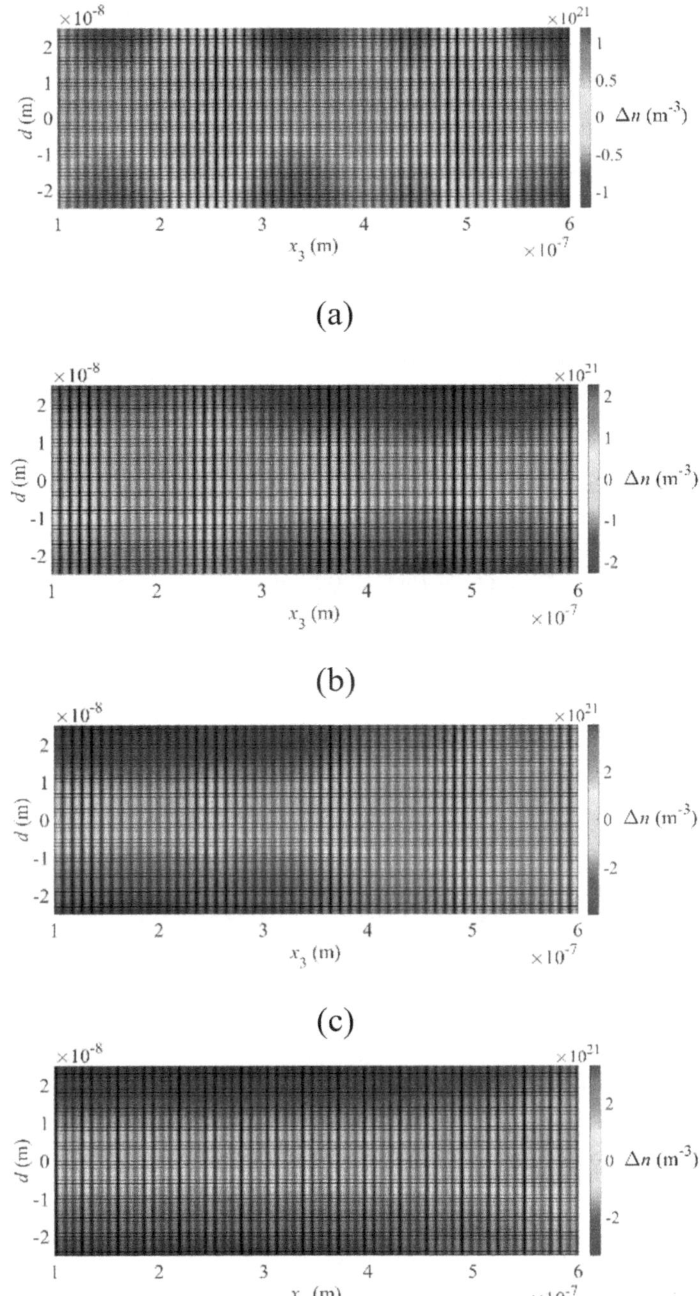

(a)

(b)

(c)

(d)

Fig. 4.10 Electron concentration perturbation $\Delta n = x_2 n^{(1)}(x_3, t)$ in $L/6 < x_3 < L$ at different time instants. (**a**) $t = 1.4$ ns. (**b**) $t = 2.8$ ns. (**c**) $t = 4.2$ ns. (**d**) $t = 5.6$ ns similar. (a) shows the electron concentration perturbation at an early instant. Then, approximately at the largest deflection in (d), the electrons form a distribution that does not change much along the beam except near the right end. Δn is linear in x_2, with more electrons at the top surface and less at the bottom.

4.6 Buckling

Consider a ZnO beam with a circular cross section as shown in Fig. 4.11. It is under a compressive axial force P. We study the buckling of the beam in the y-z plane [5]. Initially P is applied while the two ends are electrically shorted. P produces a compressive axial force only without an axial electric field and carrier concentration perturbation. Then the compressed beam is disturbed for its buckling when the two ends are electrically open.

We consider an n-type piezoelectric semiconductor beam in static bending without shear deformation. From the relevant equations in Sect. 4.2, we have

$$\begin{aligned} &Q_{,3} - Pv_{,33} = 0, \\ &D_{3,3}^{(1)} - D_2^{(0)} = qI\left(-n^{(1)}\right), \\ &J_{3,3}^{n(1)} - J_2^{n(0)} = 0, \end{aligned} \tag{4.81}$$

$$M = -\bar{c}_{33}Iv_{,33} + \bar{e}_{33}I\phi_{,3}^{(1)}, \tag{4.82}$$

$$Q = M_{,3} = -\bar{c}_{33}Iv_{,333} + \bar{e}_{33}I\phi_{,33}^{(1)}, \tag{4.83}$$

$$\begin{aligned} D_2^{(0)} &= -\bar{\varepsilon}_{11}A\phi^{(1)}, \\ D_3^{(1)} &= -\bar{e}_{33}Iv_{,33} - \bar{\varepsilon}_{33}I\phi_{,3}^{(1)}, \end{aligned} \tag{4.84}$$

$$\begin{aligned} J_2^{n(0)} &= -qn_0\mu_{11}^n A\phi^{(1)} + qD_{11}^n An^{(1)}, \\ J_3^{n(1)} &= -qn_0\mu_{33}^n I\phi_{,3}^{(1)} + qD_{33}^n In_{,3}^{(1)}, \end{aligned} \tag{4.85}$$

where, in Eq. $(4.81)_1$, we have added a term $-Pv_{,33}$ [6] which describes the contribution of the axial force to bending and is responsible for buckling. The substitution of Eqs. (4.83), (4.84), and (4.85) into Eq. (4.81) yields the following three equations for v, $\phi^{(1)}$, and $n^{(1)}$:

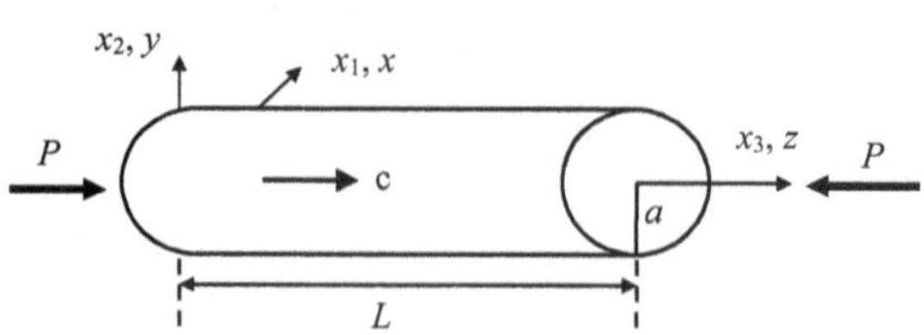

Fig. 4.11 A ZnO beam in compression

$$-\bar{c}_{33}Iv_{,3333}+\bar{e}_{33}I\phi^{(1)}_{,333}-Pv_{,33}=0,\tag{4.86}$$

$$-\bar{e}_{33}Iv_{,333}-\bar{\varepsilon}_{33}I\phi^{(1)}_{,33}+\bar{\varepsilon}_{11}A\phi^{(1)}=qI\left(-n^{(1)}\right),\tag{4.87}$$

$$-n_0\mu^n_{33}I\phi^{(1)}_{,33}+D^n_{33}In^{(1)}_{,33}+n_0\mu^n_{11}A\phi^{(1)}-D^n_{11}An^{(1)}=0.\tag{4.88}$$

We consider a simply-supported beam with zero deflection and zero bending moment at both ends, which is the simplest case for buckling analysis. The boundary conditions are that at $x_3=0$ and L:

$$\begin{aligned}
&v=0,\\
&M=-\bar{c}_{33}Iv_{,33}+\bar{e}_{33}I\phi^{(1)}_{,3}=0,\\
&D^{(1)}_3=-\bar{e}_{33}Iv_{,33}-\bar{\varepsilon}_{33}I\phi^{(1)}_{,3}=0,\\
&J^{n(1)}_3=-qn_0\mu^n_{33}I\phi^{(1)}_{,3}+qD^n_{33}In^{(1)}_{,3}=0.
\end{aligned}\tag{4.89}$$

We need to solve Eqs. (4.86), (4.87), and (4.88) under Eq. (4.89). Let

$$\begin{aligned}
&v=A\sin\frac{m\pi}{L}x_3,\quad \phi^{(1)}=B\cos\frac{m\pi}{L}x_3,\\
&n^{(1)}=C\cos\frac{m\pi}{L}x_3,\quad m=1,2,3\cdots,
\end{aligned}\tag{4.90}$$

where A, B, and C are undetermined constants. Equation (4.90) satisfies Eq. (4.89). Substituting Eq. (4.90) into Eqs. (4.86), (4.87), and (4.88), we obtain three linear homogeneous equations for A, B, and C. For nontrivial solutions, the determinant of the coefficient matrix of the linear equations has to vanish, which leads to an equation for P that determines the critical buckling load denoted by P_{cr}. The corresponding nontrivial solutions of A, B, and C determine the buckling modes for the deflection, electric potential, and carrier concentration perturbation.

The buckling load P_{cr} is found to be

$$\frac{P_{cr}}{P^e_{cr}}=1+\frac{\bar{k}^2_{33}}{1+\frac{\bar{\varepsilon}_{11}A}{\bar{\varepsilon}_{33}I}\left(\frac{L}{m\pi}\right)^2+\frac{qn_0\mu^n_{33}}{\bar{\varepsilon}_{33}D^n_{33}}\left(\frac{L}{m\pi}\right)^2},\tag{4.91}$$

where we have denoted

$$P^e_{cr}=\bar{c}_{33}I\left(\frac{m\pi}{L}\right)^2,\quad \bar{k}^2_{33}=\frac{\bar{e}^2_{33}}{\bar{\varepsilon}_{33}\bar{c}_{33}}.\tag{4.92}$$

P^e_{cr} is the buckling load of the corresponding elastic dielectric beam when electromechanical couplings are neglected. $\bar{k}^2_{33}$ is the axial electromechanical coupling factor. Its role in Eq. (4.91) represents the piezoelectric stiffening effect which tends to increase the buckling load. The effect of semiconduction represented by the n_0 term in the denominator of Eq. (4.91) tends to reduce the piezoelectric stiffening because of the screening effect of the electrons, and hence lowers the buckling load. When n_0 is set to zero in Eq. (4.91), we have following critical buckling load P^p_{cr} of the corresponding piezoelectric dielectric beam:

$$\frac{P^p_{cr}}{P^e_{cr}} = 1 + \frac{\overline{k}^2_{33}}{1 + \frac{\overline{\varepsilon}_{11} A}{\overline{\varepsilon}_{33} I}\left(\frac{L}{m\pi}\right)^2}. \tag{4.93}$$

As a numerical example, consider a beam of radius $a = 25$ nm, length $L = 600$ nm, and initial carrier concentration $n_0 = 10^{23}/\text{m}^3$. We are mainly interested in the carrier concentration perturbation Δn, which is normalized according to

$$\overline{x}_3 = x_3/L, \quad \overline{x}_2 = x_2/a, \quad \overline{\Delta n} = \overline{x}_2 \cos m\pi\overline{x}_3, \tag{4.94}$$

and is shown in Fig. 4.12 for $m = 1$, 2, and 3. It can be seen that the electrons move toward the top and bottom (the lateral surface) of the beam during buckling and that

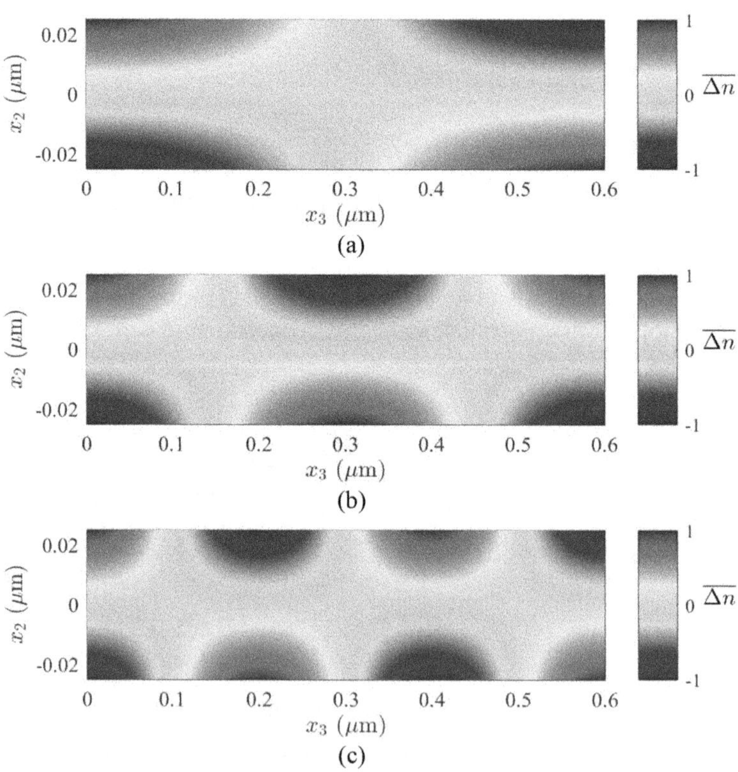

Fig. 4.12 Buckling modes of normalized Δn. (**a**) $m = 1$. (**b**) $m = 2$. (**c**) $m = 3$

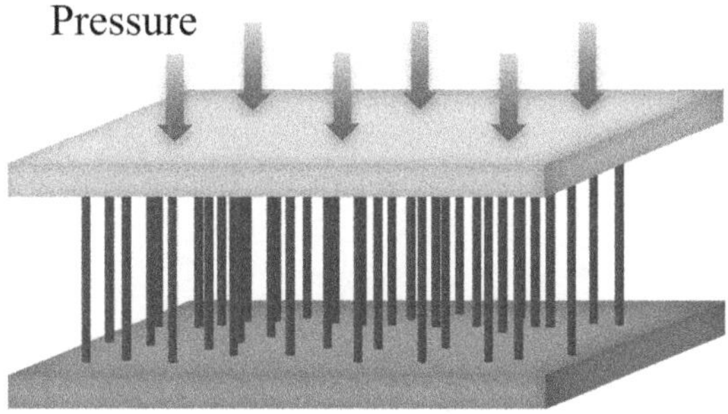

Fig. 4.13 A beam array under pressure

Δn alternates its sign across nodal points (zeros) when $m = 2$ and 3. The average of Δn over a cross section is zero. Therefore this buckling-induced Δn does not contribute directly to the axial flow of electrons in a device operating with the extensional deformation of the beam, but it does produce a lateral electric field that acts on the electrons. The distribution of the normalized φ looks the same as Δn and is not shown.

For beams with other end supports, the critical buckling loads can be determined [5] using the concept of the effective length of a beam according to the mechanics of materials [7]. Piezoelectric semiconductor beam devices often use beam arrays between two films (see Fig. 4.13), which usually operate under a pressure. The buckling load for a single beam in Eq. (4.91) can be used to determine the buckling pressure of an array [5].

References

1. Z.L. Wang, *Piezotronics and Piezo-Phototronics* (Science Press, Beijing, 2012). in Chinese
2. C.L. Zhang, X.Y. Wang, W.Q. Chen, J.S. Yang, Bending of a cantilever piezoelectric semiconductor fiber under an end force, in *Generalized Models and Non-Classical Approaches in Complex Materials*, ed. by H. Altenbach et al., (Springer, Cham, 2018), pp. 261–278
3. X.Y. Dai, F. Zhu, Z.H. Qian, J.S. Yang, Electric potential and carrier distribution in a piezoelectric semiconductor nanowire in time-harmonic bending vibration. Nano Energy **43**, 22–28 (2018)
4. Y.X. Liang, Y.T. Hu, J.S. Yang, Transient bending vibration of a piezoelectric semiconductor nanofiber under a suddenly applied shear force. Acta Mech. Solida Sin. **32**, 688–697 (2019)
5. C. Liang, C.L. Zhang, W.Q. Chen, J.S. Yang, Static buckling of piezoelectric semiconductor fibers. Mater. Res. Express **6**, 125919 (2020)
6. J.S. Yang, Equations for the extension and flexure of a piezoelectric beam with rectangular cross section and applications. Int. J. Appl. Electromagn. Mech. **9**, 409–420 (1998)
7. J.M. Gere, S.P. Timoshenko, *Mechanics of Materials*, 2nd edn. (Wadsworth, Belmont, 1984)

Chapter 5
Extension and Bending of Plates

In addition to beams or fibers, piezoelectric semiconductor plates or thin films are also widely used in devices in extension and bending. In this chapter, we derive two-dimensional equations for thin plates of piezoelectric semiconductors [1, 2]. We follow R.D. Mindlin's approach [3–5], beginning from the three-dimensional equations with power series expansions in the plate thickness coordinate and then truncating the expansions to obtain low-order equations for extension and/or bending. The last section is on shells.

5.1 Recapitulation of Three-Dimensional Equations

Different from rods and beams for which the piezoelectric constitutive relations in Eq. (1.3) are convenient, for plates we use Eq. $(1.2)_{1,2}$ most of the time. For convenience we gather the complete set of the linearized three-dimensional equation below from Eqs. $(1.1)_1$ with $\mathbf{f} = 0$, (1.10), $(1.2)_{1,2}$, (1.9) for uniform p_0 and n_0, and (1.5):

$$\begin{aligned} T_{ij,i} &= \rho \ddot{u}_j, \\ D_{i,i} &= q(\Delta p - \Delta n), \end{aligned} \tag{5.1}$$

$$\begin{aligned} J^p_{i,i} &= -q\frac{\partial}{\partial t}(\Delta p), \\ J^n_{i,i} &= q\frac{\partial}{\partial t}(\Delta n), \end{aligned} \tag{5.2}$$

$$\begin{aligned} T_{ij} &= c_{ijkl}S_{kl} - e_{kij}E_k, \\ D_i &= e_{ikl}S_{kl} + \varepsilon_{ik}E_k, \end{aligned} \tag{5.3}$$

J. Yang, *Analysis of Piezoelectric Semiconductor Structures*,
https://doi.org/10.1007/978-3-030-48206-0_5

$$J_i^p = qp_0\mu_{ij}^p E_j - qD_{ij}^p P_j,$$
$$J_i^n = qn_0\mu_{ij}^n E_j + qD_{ij}^n N_j, \tag{5.4}$$

$$S_{ij} = (u_{i,j} + u_{j,i})/2,$$
$$E_i = -\varphi_{,i}, \tag{5.5}$$

where we have denoted the carrier concentration perturbation gradients by

$$P_j = (\Delta p)_{,j}, \quad N_j = (\Delta n)_{,j}. \tag{5.6}$$

From Eqs. (1.8) and (1.7), we have

$$p_0 = N_A^-, \quad n_0 = N_D^+, \tag{5.7}$$

which are constants for uniform doping, and

$$p = p_0 + \Delta p, \quad n = n_0 + \Delta n. \tag{5.8}$$

5.2 Hierarchy of Two-Dimensional Equations

Consider the plate of uniform thickness $2h$ shown in Fig. 5.1. The x_1 and x_2 axes are in the plate middle plane where $x_3 = 0$. The metal electrodes are assumed to be very thin with negligible mechanical effects.

We expand the mechanical displacement vector, the electric potential, and the carrier concentration perturbations into power series in x_3:

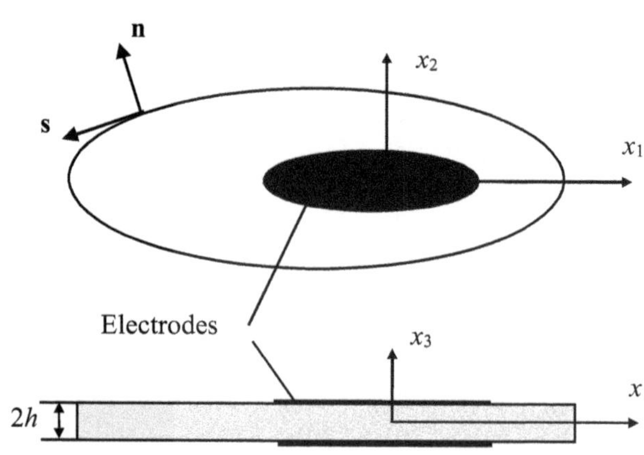

Fig. 5.1 Plan view and cross section of a thin piezoelectric semiconductor plate

$$
\begin{aligned}
u_i(x_1,x_2,x_3,t) &= \sum_{m=0}^{\infty} x_3^m u_i^{(m)}(x_1,x_2,t),\\
\varphi(x_1,x_2,x_3,t) &= \sum_{m=0}^{\infty} x_3^m \varphi^{(m)}(x_1,x_2,t),
\end{aligned}
\tag{5.9}
$$

$$
\begin{aligned}
\Delta p(x_1,x_2,x_3,t) &= \sum_{m=0}^{\infty} x_3^m p^{(m)}(x_1,x_2,t),\\
\Delta n(x_1,x_2,x_3,t) &= \sum_{m=0}^{\infty} x_3^m n^{(m)}(x_1,x_2,t).
\end{aligned}
\tag{5.10}
$$

Our goal is to obtain two-dimensional equations for $u_i^{(m)}$, $\varphi^{(m)}$, $p^{(m)}$ and $n^{(m)}$. Corresponding to Eqs. (5.9) and (5.10), from Eqs. (5.5) and (5.6), we obtain

$$
S_{ij} = \sum_{m=0}^{\infty} x_3^m S_{ij}^{(m)}, \quad E_i = \sum_{m=0}^{\infty} x_3^m E_i^{(m)}, \tag{5.11}
$$

$$
P_i = \sum_{m=0}^{\infty} x_3^m P_i^{(m)}, \quad N_i = \sum_{m=0}^{\infty} x_3^m N_i^{(m)}, \tag{5.12}
$$

where

$$
\begin{aligned}
S_{ij}^{(m)} &= \frac{1}{2}\left[u_{j,i}^{(m)} + u_{i,j}^{(m)} + (m+1)\left(\delta_{i3}u_j^{(m+1)} + \delta_{3j}u_i^{(m+1)}\right)\right],\\
E_i^{(m)} &= -\varphi_{,i}^{(m)} - (m+1)\delta_{3i}\varphi^{(m+1)},
\end{aligned}
\tag{5.13}
$$

$$
\begin{aligned}
P_j^{(m)} &= p_{,j}^{(m)} + (m+1)\delta_{3j}p^{(m+1)},\\
N_j^{(m)} &= n_{,j}^{(m)} + (m+1)\delta_{3j}n^{(m+1)}.
\end{aligned}
\tag{5.14}
$$

Then we multiply Eqs. (5.1) and (5.2) by x_3^m and integrate the equations through the plate thickness. This gives the following plate equations of motion, electrostatics, and continuity:

$$
\begin{aligned}
T_{aj,a}^{(m)} - mT_{3j}^{(m-1)} + t_j^{(m)} &= \sum_{n=0}^{\infty} \rho B^{(m,n)} \ddot{u}_j^{(n)}, \quad a = 1,2,\\
D_{a,a}^{(m)} - mD_3^{(m-1)} + d^{(m)} &= \sum_{n=0}^{\infty} qB^{(m,n)}\left(p^{(n)} - n^{(n)}\right),
\end{aligned}
\tag{5.15}
$$

$$
\begin{aligned}
J_{a,a}^{p(m)} - mJ_3^{p(m-1)} + j^{p(m)} &= -\sum_{n=0}^{\infty} qB^{(m,n)}\dot{p}^{(n)},\\
J_{a,a}^{n(m)} - mJ_3^{n(m-1)} + j^{n(m)} &= \sum_{n=0}^{\infty} qB^{(m,n)}\dot{n}^{(n)},
\end{aligned}
\tag{5.16}
$$

where we have introduced an index convention that subscripts a, b, c, and d assume 1 and 2 only but not 3. In Eqs. (5.15) and (5.16), the plate electromechanical resultants are defined by the following integration along the plate thickness:

$$\left\{T_{ij}^{(m)}, D_i^{(m)}, J_i^{p(m)}, J_i^{n(m)}\right\} = \int_{-h}^{h} x_3^m \{T_{ij}, D_i, J_i^p, J_i^n\} dx_3. \tag{5.17}$$

The electromechanical loads at the plate top and bottom surfaces are represented by

$$\begin{aligned} t_j^{(m)} &= \left[x_3^m T_{3j}\right]_{-h}^{h}, \quad d^{(m)} = \left[x_3^m D_3\right]_{-h}^{h}, \\ j^{p(m)} &= \left[x_3^m J_3^p\right]_{-h}^{h}, \quad j^{n(m)} = \left[x_3^m J_3^n\right]_{-h}^{h}. \end{aligned} \tag{5.18}$$

In addition,

$$B^{(m,n)} = \int_{-h}^{h} x_3^m x_3^n dx_3 = \begin{cases} 2h^{m+n+1}/(m+n+1), & m+n \quad \text{even}, \\ 0, \quad m+n & \text{odd}. \end{cases} \tag{5.19}$$

Next we derive plate constitutive equations. Substituting Eqs. (5.3) and (5.4) into Eq. (5.17), using Eqs. (5.11), (5.12), and (5.19), we obtain the plate constitutive equations as follows:

$$\begin{aligned} T_{ij}^{(m)} &= \sum_{n=0}^{\infty} B^{(m,n)} \left(c_{ijkl} S_{kl}^{(n)} - e_{kij} E_k^{(n)}\right), \\ D_i^{(m)} &= \sum_{n=0}^{\infty} B^{(m,n)} \left(e_{ijk} S_{jk}^{(n)} + \varepsilon_{ij} E_j^{(n)}\right), \end{aligned} \tag{5.20}$$

$$\begin{aligned} J_i^{p(m)} &= \sum_{n=0}^{\infty} B^{(m,n)} \left(\mu_{ij}^p E_j^{(n)} - D_{ij}^p P_j^{(n)}\right), \\ J_i^{n(m)} &= \sum_{n=0}^{\infty} B^{(m,n)} \left(\mu_{ij}^n E_j^{(n)} + D_{ij}^n N_j^{(n)}\right). \end{aligned} \tag{5.21}$$

With successive substitutions from Eqs. (5.20), (5.21), (5.13), and (5.14), we can write Eqs. (5.15) and (5.16) as equations for $u_i^{(n)}$, $\varphi^{(n)}$, $p^{(n)}$, and $n^{(n)}$.

5.3 Zero- and First-Order Equations

In this section we truncate the hierarchy of two-dimensional equations in the previous section to zero- and first-order equations. These low-order equations are not ready to be used for analyzing plate extension or bending problems. Instead, they will be used to derive equations for extension or bending in subsequent sections with further modifications of constitutive relations. For a complete description of zero- and first-order strains, electric fields, and carrier concentration perturbations, we truncate the expansions in the previous section as follows:

$$\begin{aligned} u_j(x_1,x_2,x_3,t) &\cong u_j^{(0)}(x_1,x_2,t) \\ &\quad + x_3 u_j^{(1)}(x_1,x_2,t) + x_3^2 u_j^{(2)}(x_1,x_2,t), \\ \varphi(x_1,x_2,x_3,t) &\cong \varphi^{(0)}(x_1,x_2,t) \\ &\quad + x_3 \varphi^{(1)}(x_1,x_2,t) + x_3^2 \varphi^{(2)}(x_1,x_2,t), \end{aligned} \tag{5.22}$$

$$\begin{aligned} \Delta p(x_1,x_2,x_3,t) &\cong p^{(0)}(x_1,x_2,t) \\ &\quad + x_3 p^{(1)}(x_1,x_2,t) + x_3^2 p^{(2)}(x_1,x_2,t), \\ \Delta n(x_1,x_2,x_3,t) &\cong n^{(0)}(x_1,x_2,t) \\ &\quad + x_3 n^{(1)}(x_1,x_2,t) + x_3^2 n^{(2)}(x_1,x_2,t), \end{aligned} \tag{5.23}$$

where $u_a^{(0)}$ ($a = 1$ and 2) are the extensional displacements, $u_a^{(1)}$ the fundamental thickness-shear displacements, $u_a^{(2)}$ the second-order or symmetric thickness-shear displacements, $u_3^{(0)}$ the bending or flexural displacement, $u_3^{(1)}$ the fundamental thickness-stretch displacement, and $u_3^{(2)}$ the second-order thickness-stretch displacement. The corresponding strains, electric field and carrier concentration gradients are

$$\begin{aligned} S_{ij} &\cong S_{ij}^{(0)} + x_3 S_{ij}^{(1)}, \quad E_i \cong E_i^{(0)} + x_3 E_i^{(1)}, \\ P_i &\cong P_i^{(0)} + x_3 P_i^{(1)}, \quad N_i \cong N_i^{(0)} + x_3 N_i^{(1)}, \end{aligned} \tag{5.24}$$

where

$$\begin{aligned} &S_{11}^{(0)} = u_{1,1}^{(0)}, \quad S_{22}^{(0)} = u_{2,2}^{(0)}, \quad S_{33}^{(0)} = u_3^{(1)} \\ &2S_{23}^{(0)} = u_{3,2}^{(0)} + u_2^{(1)}, \quad 2S_{31}^{(0)} = u_{3,1}^{(0)} + u_1^{(1)}, \quad 2S_{12}^{(0)} = u_{1,2}^{(0)} + u_{2,1}^{(0)}, \end{aligned} \tag{5.25}$$

$$\begin{aligned} &S_{11}^{(1)} = u_{1,1}^{(1)}, \quad S_{22}^{(1)} = u_{2,2}^{(1)}, \quad S_{33}^{(1)} = 2u_3^{(2)}, \\ &2S_{23}^{(1)} = u_{3,2}^{(1)} + 2u_2^{(2)}, \quad 2S_{31}^{(2)} = u_{3,1}^{(1)} + 2u_1^{(2)}, \quad 2S_{12}^{(1)} = u_{1,2}^{(1)} + u_{2,1}^{(1)}, \end{aligned} \tag{5.26}$$

$$E_1^{(0)} = -\varphi_{,1}^{(0)}, \quad E_2^{(0)} = -\varphi_{,2}^{(0)}, \quad E_3^{(0)} = -\varphi^{(1)}, \tag{5.27}$$

$$E_1^{(1)} = -\varphi_{,1}^{(1)}, \quad E_2^{(1)} = -\varphi_{,2}^{(1)}, \quad E_3^{(1)} = -2\varphi^{(2)}, \tag{5.28}$$

$$P_1^{(0)} = p_{,1}^{(0)}, \quad P_2^{(0)} = p_{,2}^{(0)}, \quad P_3^{(0)} = p^{(1)}, \tag{5.29}$$

$$P_1^{(1)} = p_{,1}^{(1)}, \quad P_2^{(1)} = p_{,2}^{(1)}, \quad P_3^{(1)} = 2p^{(2)}, \tag{5.30}$$

$$N_1^{(0)} = n_{,1}^{(0)}, \quad N_2^{(0)} = n_{,2}^{(0)}, \quad N_3^{(0)} = n^{(1)}, \tag{5.31}$$

$$N_1^{(1)} = n_{,1}^{(1)}, \quad N_2^{(1)} = n_{,2}^{(1)}, \quad N_3^{(1)} = 2n^{(2)}. \tag{5.32}$$

The zero- and first-order equations of motion, the charge equation of electrostatics, and the continuity equations for holes and electrons are

$$\begin{aligned} &T_{aj,a}^{(0)} + t_j^{(0)} = 2h\rho\ddot{u}_j^{(0)}, \\ &T_{aj,a}^{(1)} - T_{3j}^{(0)} + t_j^{(1)} = \frac{2h^3\rho}{3}\ddot{u}_j^{(1)}, \end{aligned} \tag{5.33}$$

$$\begin{aligned} &D_{a,a}^{(0)} + d^{(0)} = 2hq\left(p^{(0)} - n^{(0)}\right), \\ &D_{a,a}^{(1)} - D_3^{(0)} + d^{(1)} = \frac{2h^3}{3}q\left(p^{(1)} - n^{(1)}\right), \end{aligned} \tag{5.34}$$

$$\begin{aligned} &J_{a,a}^{p(0)} + j^{p(0)} = -2hq\dot{p}^{(0)}, \\ &J_{a,a}^{p(1)} - J_3^{p(0)} + j^{p(1)} = -\frac{2h^3}{3}q\dot{p}^{(1)}, \end{aligned} \tag{5.35}$$

$$\begin{aligned} &J_{a,a}^{n(0)} + j^{n(0)} = 2hq\dot{n}^{(0)}, \\ &J_{a,a}^{n(1)} - J_3^{n(0)} + j^{n(1)} = \frac{2h^3}{3}q\dot{n}^{(1)}, \end{aligned} \tag{5.36}$$

where $B^{(0,0)} = 2\,h$ and $B^{(1,1)} = 2\,h^3/3$ have been used. The plate constitutive relations are

$$\begin{aligned} &T_{ij}^{(0)} = 2h\left(c_{ijkl}S_{kl}^{(0)} - e_{kij}E_k^{(0)}\right), \\ &D_i^{(0)} = 2h\left(e_{ijk}S_{jk}^{(0)} + \varepsilon_{ij}E_j^{(0)}\right), \end{aligned} \tag{5.37}$$

$$\begin{aligned} &J_i^{p(0)} = 2h\left(\mu_{ij}^p E_j^{(0)} - D_{ij}^p P_j^{(0)}\right), \\ &J_i^{n(0)} = 2h\left(\mu_{ij}^n E_j^{(0)} + D_{ij}^n N_j^{(0)}\right), \end{aligned} \tag{5.38}$$

$$T_{ij}^{(1)} = \frac{2h^3}{3}\left(c_{ijkl}S_{kl}^{(1)} - e_{kij}E_k^{(1)}\right),$$
$$D_i^{(1)} = \frac{2h^3}{3}\left(e_{ijk}S_{jk}^{(1)} + \varepsilon_{ij}E_j^{(1)}\right), \tag{5.39}$$

$$J_i^{p(1)} = \frac{2h^3}{3}\left(\mu_{ij}^p E_j^{(1)} - D_{ij}^p P_j^{(1)}\right),$$
$$J_i^{n(1)} = \frac{2h^3}{3}\left(\mu_{ij}^n E_j^{(1)} + D_{ij}^n N_j^{(1)}\right), \tag{5.40}$$

where $T_{ab}^{(0)}$ represent the plate in-plane extensional and shear forces, $T_{b3}^{(0)}$ the transverse shear forces in bending, and $T_{ab}^{(1)}$ the bending and twisting moments. The zero- and first-order surface loads are

$$t_j^{(0)} = T_{3j}(h) - T_{3j}(-h), \quad d^{(0)} = D_3(h) - D_3(-h),$$
$$j^{p(0)} = J_3^p(h) - J_3^p(-h), \quad j^{n(0)} = J_3^n(h) - J_3^n(-h), \tag{5.41}$$

$$t_j^{(1)} = h\left[T_{3j}(h) + T_{3j}(-h)\right], \quad d^{(1)} = h[D_3(h) + D_3(-h)],$$
$$j^{p(1)} = h\left[J_3^p(h) + J_3^p(-h)\right], \quad j^{n(1)} = h\left[J_3^n(h) + J_3^n(-h)\right]. \tag{5.42}$$

5.4 Equations for Extension

In this section, we reduce the zero-order equations in the previous section to equations for extension. This is mainly a treatment of the mechanical fields and equations. The electrical fields and equations are not reduced. Both the zero- and first-order electrical fields are kept so that they are available in different situations for electroded or unelectroded plates. For the extensional motion of a thin plate, the relevant in-plane extensional displacements are $u_b^{(0)}$ where $b = 1$ and 2. The in-plane extensional and shear stresses as well as strains are approximately uniform along the plate thickness. The zero-order equations of motion and constitutive relations are sufficient:

$$T_{ab,a}^{(0)} + t_b^{(0)} = 2h\rho\ddot{u}_b^{(0)}, \quad a, b = 1, 2, \tag{5.43}$$

$$T_{ij}^{(0)} = 2h\left(c_{ijkl}S_{kl}^{(0)} - e_{kij}E_k^{(0)}\right),$$
$$D_i^{(0)} = 2h\left(e_{ikl}S_{kl}^{(0)} + \varepsilon_{ij}E_j^{(0)}\right). \tag{5.44}$$

From Eq. (5.25), it can be seen that $S_{33}^{(0)}$ is involved with $u_3^{(1)}$, and it describes the Poisson's effect in extension. $S_{31}^{(0)}$ and $S_{32}^{(0)}$ are involved with $u_a^{(1)}$ and $u_3^{(0)}$. They describe shear deformation in bending but may be present in the extensional constitutive relations in Eq. (5.44) because of anisotropy. Therefore Eq. (5.44) is not ready to be used for extension yet. We take Poisson's effect into consideration by the usual plate thickness stress relaxation condition $T_{33}^{(0)} = 0$. We also relaxed the plate shear stresses associated with bending by setting $T_{31}^{(0)} = 0$ and $T_{32}^{(0)} = 0$. In summary, we take the following to be approximately true:

$$T_{3j}^{(0)} = 0, \quad j = 1, 2, 3. \tag{5.45}$$

Equation (5.45) is the usual stress relaxation for extension. According to the compact matrix notation, with the range of p and q as 1, 2, ... and 6, Eq. (5.45) can be written as

$$T_{q}^{(0)} = 0, \quad q = 3, 4, 5. \tag{5.46}$$

For convenience ,we introduce another index convention in which subscripts u, v, and w take the values 3, 4, and 5 while subscripts r and s take the remaining values 1, 2, and 6. Then Eq. (5.44) can be written as

$$\begin{aligned}
T_{r}^{(0)} &= 2h\left(c_{rs}S_{s}^{(0)} + c_{ru}S_{u}^{(0)} - e_{kr}E_{k}^{(0)}\right),\\
T_{v}^{(0)} &= 2h\left(c_{vs}S_{s}^{(0)} + c_{vw}S_{w}^{(0)} - e_{kv}E_{k}^{(0)}\right) = 0,\\
D_{i}^{(0)} &= 2h\left(e_{is}S_{s}^{(0)} + e_{iu}S_{u}^{(0)} - \varepsilon_{ij}E_{j}^{(0)}\right),
\end{aligned} \tag{5.47}$$

where Eq. (5.46) has been used. From Eq. $(5.47)_2$ we solve for

$$S_{u}^{(0)} = -c_{uv}^{-1}c_{vs}S_{s}^{(0)} + c_{uv}^{-1}e_{kv}E_{k}^{(0)}. \tag{5.48}$$

The substitution of Eq. (5.48) into Eq. $(5.47)_{1,3}$ gives

$$\begin{aligned}
T_{r}^{(0)} &= 2h\left(\gamma_{rs}S_{s}^{(0)} - \psi_{kr}E_{k}^{(0)}\right),\\
D_{i}^{(0)} &= 2h\left(\psi_{is}S_{s}^{(0)} + \zeta_{ij}E_{j}^{(0)}\right),
\end{aligned} \tag{5.49}$$

where the effective plate material constants relaxed for the extension of thin plates are

$$\begin{aligned}
\gamma_{rs} &= c_{rs} - c_{rv}c_{vw}^{-1}c_{ws}, \quad r, s = 1, 2, 6,\\
\psi_{ks} &= e_{ks} - e_{kw}c_{wv}^{-1}c_{vs}, \quad v, w = 3, 4, 5,\\
\zeta_{kj} &= \varepsilon_{kj} + e_{kv}c_{vw}^{-1}e_{jw}, \quad j, k = 1, 2, 3.
\end{aligned} \tag{5.50}$$

We note that γ_{rs} are Voigt's anisotropic plate elastic constants. Equation (5.49) is the constitutive relation suitable for extension. Its right-hand side does not have $S_{3j}^{(0)}$ and hence are not involved with $u_a^{(1)}$ and $u_3^{(0)}$. When Eq. $(5.49)_1$ is substituted into Eq. (5.43), mechanically it results in equations involving the extensional displacement $u_a^{(0)}$ only. At the boundary of a plate edge with in-plane unit exterior normal **n** and in-plane unit tangent **s** as shown in Fig. 5.1, the mechanical boundary conditions for extension may be the prescription of

$$T_{nn}^{(0)} \quad \text{or} \quad u_n^{(0)}, \quad \text{and} \quad T_{ns}^{(0)} \quad \text{or} \quad u_s^{(0)}. \tag{5.51}$$

5.5 Equations for Bending

Similar to the previous section, in this section we also mainly focus on the reduction of the mechanical equations and will leave the electrical equations untouched. For bending of plates with shear deformation, we want to obtain equations for the bending displacement $u_3^{(0)}$ and fundamental thickness-shear displacements $u_a^{(1)}$. However, in materials with strong anisotropy, $u_a^{(1)}$ may be coupled to extension. This coupling is kept in the following, which may be eliminated through proper stress relaxation when needed or simply neglected as an approximation. In bending, the in-plane normal and shear stresses vary linearly along the plate thickness to produced bending and twisting moments. Accordingly, we keep the first-order strains as follows:

$$S_{ij} \cong S_{ij}^{(0)} + x_3 S_{ij}^{(1)}. \tag{5.52}$$

From Eqs. (5.25) and (5.26), it can be seen that $S_{33}^{(0)}$ is involved with $u_3^{(1)}$, and $S_{3j}^{(1)}$ are involved with $u_j^{(2)}$. These higher-order plate displacements will be eliminated through two separate stress relaxation procedures in the following for the zero- and first-order constitutive relations, respectively.

For the zero-order constitutive relations, from Eq. $(5.37)_1$, by setting $i = j = 3$, we set

$$T_{33}^{(0)} = 2h\left(c_{33kl}S_{kl}^{(0)} - e_{k33}E_k^{(0)}\right) = 0. \tag{5.53}$$

This implies the following expression for $S_{33}^{(0)}$:

$$S_{33}^{(0)} = -\frac{1}{c_{3333}}\left(c_{33kl}S_{kl}^{(0)} - c_{3333}S_{33}^{(0)} - e_{k33}E_k^{(0)}\right). \tag{5.54}$$

In Eq. (5.54), $S_{33}^{(0)}$ has been eliminated on the right-hand side because when $i = j = 3$, the two terms containing $S_{33}^{(0)}$ cancel with each other. From Eq. (5.54), the thickness expansion or contraction accompanying the in-plane extension of the plate due to Poisson's effect can be found if interested. Substituting Eq. (5.54) back into Eq. (5.37), we obtain

$$\begin{aligned} T_{ij}^{(0)} &= 2h\left(\bar{c}_{ijkl}S_{kl}^{(0)} - \bar{e}_{kij}E_k^{(0)}\right), \\ D_i^{(0)} &= 2h\left(\bar{e}_{ikl}S_{kl}^{(0)} + \bar{\varepsilon}_{ij}E_j^{(0)}\right), \end{aligned} \tag{5.55}$$

where the relaxed material constants are defined by

$$\begin{aligned} \bar{c}_{ijkl} &= c_{ijkl} - c_{ij33}c_{33kl}/c_{3333}, \\ \bar{e}_{kij} &= e_{kij} - e_{k33}c_{33ij}/c_{3333}, \\ \bar{\varepsilon}_{ij} &= \varepsilon_{ij} + e_{i33}e_{j33}/c_{3333}. \end{aligned} \tag{5.56}$$

We note that the right-hand side of Eq. (5.55) does not depend on $S_{33}^{(0)}$ and $T_{33}^{(0)} = 0$ is automatically satisfied by Eq. (5.55). However, for the transverse shear forces in bending, the right-hand side of $T_{3b}^{(0)}$ from Eq. (5.55) may depend on the extensional strains $S_{ab}^{(0)}$ and hence the extensional displacements $u_a^{(0)}$ in materials with high anisotropy.

Since we are using a linear function $x_3 u_a^{(1)}$ as an approximation for the fundamental thickness-shear deformation which in dynamic problems has a sinusoidal variation along the plate thickness, we introduce two shear correction factors κ_1 and κ_2 to improve the accuracy of the plate equations by modifying the following zero-order plate shear strains:

$$S_{31}^{(0)} \to \kappa_1 S_{31}^{(0)}, \quad S_{32}^{(0)} \to \kappa_2 S_{32}^{(0)}. \tag{5.57}$$

This modifies the $\bar{c}_{ijkl}$ and $\bar{e}_{ijk}$ in Eq. (5.55) into $c_{ijkl}^{(0)}$ and $e_{ijk}^{(0)}$. Then Eq. (5.55) takes the following form in the compact matrix notation:

$$\begin{aligned} T_p^{(0)} &= 2h\left(c_{pq}^{(0)}S_q^{(0)} - e_{kp}^{(0)}E_k^{(0)}\right), \\ D_i^{(0)} &= 2h\left(e_{iq}^{(0)}S_q^{(0)} + \bar{\varepsilon}_{ij}E_j^{(0)}\right). \end{aligned} \tag{5.58}$$

The structures of the $c_{pq}^{(0)}$ and $e_{iq}^{(0)}$ matrices are [6]

$$\left[c_{pq}^{(0)}\right] = \begin{pmatrix} \bar{c}_{11} & \bar{c}_{12} & 0 & \kappa_2\bar{c}_{14} & \kappa_1\bar{c}_{15} & \bar{c}_{16} \\ \bar{c}_{21} & \bar{c}_{22} & 0 & \kappa_2\bar{c}_{24} & \kappa_1\bar{c}_{25} & \bar{c}_{26} \\ 0 & 0 & 0 & 0 & 0 & 0 \\ \kappa_2\bar{c}_{41} & \kappa_2\bar{c}_{42} & 0 & \kappa_2\kappa_2\bar{c}_{44} & \kappa_1\kappa_2\bar{c}_{45} & \kappa_2\bar{c}_{46} \\ \kappa_1\bar{c}_{51} & \kappa_1\bar{c}_{52} & 0 & \kappa_1\kappa_2\bar{c}_{54} & \kappa_1\kappa_1\bar{c}_{55} & \kappa_1\bar{c}_{56} \\ \bar{c}_{61} & \bar{c}_{62} & 0 & \kappa_2\bar{c}_{64} & \kappa_1\bar{c}_{65} & \bar{c}_{66} \end{pmatrix}, \tag{5.59}$$

$$\left[e_{iq}^{(0)}\right] = \begin{bmatrix} \bar{e}_{11} & \bar{e}_{12} & 0 & \kappa_2\bar{e}_{14} & \kappa_1\bar{e}_{15} & \bar{e}_{16} \\ \bar{e}_{21} & \bar{e}_{22} & 0 & \kappa_2\bar{e}_{24} & \kappa_1\bar{e}_{25} & \bar{e}_{26} \\ \bar{e}_{31} & \bar{e}_{32} & 0 & \kappa_2\bar{e}_{34} & \kappa_1\bar{e}_{35} & \bar{e}_{36} \end{bmatrix}. \tag{5.60}$$

The two correction factors appear as products or squares at a few places in Eq. (5.59) because they are introduced in the energy density function (the electric enthalpy) for the zero-order constitutive relations, which is a quadratic form [5, 6]. In the case of piezoelectric dielectric plates, the correction factors are determined by requiring the two fundamental thickness-shear resonance frequencies obtained from the two-dimensional plate equations to be equal to the corresponding exact frequencies predicted by the three-dimensional equations. With shear correction factors thus determined, the two-dimensional plate equations and the exact three-dimensional equations yield the same frequencies for a particular motion, i.e., the thickness-shear vibration of a plate in the two fundamental thickness-shear modes. This is useful in the analysis of plate piezoelectric devices operating with thickness-shear modes.

For the first-order constitutive relations, in Eq. $(5.39)_1$, we set

$$T_q^{(1)} = 0, \ \ q = 3,4,5. \tag{5.61}$$

Then, similar to the derivation of Eqs. (5.46) through (5.50), we obtain the relaxed first-order constitutive relations:

$$\begin{aligned} T_r^{(1)} &= \frac{2h^3}{3}\left(\gamma_{rs}S_s^{(1)} - \psi_{kr}E_k^{(1)}\right), \\ D_i^{(1)} &= \frac{2h^3}{3}\left(\psi_{is}S_s^{(1)} + \zeta_{ij}E_j^{(1)}\right), \end{aligned} \tag{5.62}$$

where the effective plate material constants γ_{rs}, ψ_{ks}, and ζ_{kj} are given by Eq. (5.50). The right-hand side of Eq. (5.62) does not depend on $S_{3j}^{(1)}$ or $u_j^{(2)}$. When the relevant ones of Eqs. $(5.58)_1$ and $(5.62)_1$ are substituted into

$$\begin{aligned} T_{a3,a}^{(0)} + t_3^{(0)} &= 2h\rho\ddot{u}_3^{(0)}, \\ T_{ab,a}^{(1)} - T_{3b}^{(0)} + t_b^{(1)} &= \frac{2h^3\rho}{3}\ddot{u}_b^{(1)}, \end{aligned} \tag{5.63}$$

we obtain three equations for the bending displacement $u_3^{(0)}$ and shear displacements $u_a^{(1)}$ $(a = 1, 2)$. In addition to the couplings to electric fields, Eq. $(5.63)_1$ may be coupled to the extensional displacements $u_a^{(0)}$ as explained earlier. At the boundary of a plate with in-plane unit exterior normal **n** and in-plane unit tangent **s** as shown in Fig. 5.1, mechanically, we may prescribe the following boundary conditions:

$$\begin{aligned} &T_{nn}^{(0)} \quad \text{or} \quad u_n^{(0)}, \quad T_{ns}^{(0)} \quad \text{or} \quad u_s^{(0)}, \quad T_{n3}^{(0)} \quad \text{or} \quad u_3^{(0)}, \\ &T_{nn}^{(1)} \quad \text{or} \quad u_n^{(1)}, \quad \text{and} \quad T_{ns}^{(1)} \quad \text{or} \quad u_s^{(1)}. \end{aligned} \tag{5.64}$$

5.6 Reduction to Bending Without Shear Deformation

For the bending of very thin plates in which the flexural motion is dominant and the shear deformation is small, the Kirchhoff classical or elementary theory of bending without shear deformation is sufficient. To reduce the theory for bending with shear deformation in the previous section to the theory of bending without shear deformation, two approximations are needed. First we take the plate thickness-shear strains $S_{a3}^{(0)}$ to vanish, which implies, from Eq. (5.25),

$$u_a^{(1)} = -u_{3,a}^{(0)} \Rightarrow S_{ab}^{(1)} = -u_{3,ab}^{(0)}. \tag{5.65}$$

Equation (5.65) enables us to express $u_a^{(1)}$ in terms of $u_3^{(0)}$ in Eqs. (5.26) and (5.62). Another approximation needed is to neglect the rotatory inertia $\rho 2h^3/3$ in Eq. $(5.63)_2$ to obtain

$$T_{ab,a}^{(1)} - T_{3b}^{(0)} + t_b^{(1)} = 0, \tag{5.66}$$

from which we can solve for $T_{3b}^{(0)}$ and substitute the result into Eq. $(5.63)_1$ to obtain the equation for elementary flexure:

$$T_{ab,ab}^{(1)} + t_{b,b}^{(1)} + t_3^{(0)} = 2h\rho \ddot{u}_3^{(0)}. \tag{5.67}$$

With substitutions from Eqs. $(5.62)_1$ and (5.65), we can write Eq. (5.67) as an equation for the bending displacement $u_3^{(0)}$. At the boundary of a plate with an in-plane unit exterior normal **n** and an in-plane unit tangent **s** (see Fig. 5.1), the boundary conditions for elementary bending without shear deformation may be

$$T_{n3}^{(0)} + T_{ns,s}^{(1)} \quad \text{or} \quad u_3^{(0)}, \quad \text{and} \quad T_{nn}^{(1)} \quad \text{or} \quad u_{3,n}^{(0)}. \tag{5.68}$$

5.7 Equations for Crystals of Class (6mm)

Plates of crystals of class (6mm) such as ZnO are widely used in devices. Under the compact matrix notation, with the c-axis along the plate thickness x_3, the material matrices of these crystals are

$$[c_{pq}] = \begin{pmatrix} c_{11} & c_{12} & c_{13} & 0 & 0 & 0 \\ c_{21} & c_{11} & c_{13} & 0 & 0 & 0 \\ c_{31} & c_{31} & c_{33} & 0 & 0 & 0 \\ 0 & 0 & 0 & c_{44} & 0 & 0 \\ 0 & 0 & 0 & 0 & c_{44} & 0 \\ 0 & 0 & 0 & 0 & 0 & c_{66} \end{pmatrix}, \tag{5.69}$$

$$[e_{ip}] = \begin{pmatrix} 0 & 0 & 0 & 0 & e_{15} & 0 \\ 0 & 0 & 0 & e_{15} & 0 & 0 \\ e_{31} & e_{31} & e_{33} & 0 & 0 & 0 \end{pmatrix}, \quad [\varepsilon_{ij}] = \begin{pmatrix} \varepsilon_{11} & 0 & 0 \\ 0 & \varepsilon_{11} & 0 \\ 0 & 0 & \varepsilon_{33} \end{pmatrix}, \tag{5.70}$$

where $c_{66} = (c_{11} - c_{12})/2$. μ_{ij} and d_{ij} have the same structure as ε_{ij}. For these materials there is no coupling between extension and shear. After the stress relaxation in Eq. (5.53), only some of the elements of Eqs. (5.69) and (5.70) are modified. Eq. (5.56) takes the following specific form:

$$[\overline{c}_{pq}] = \begin{pmatrix} \overline{c}_{11} & \overline{c}_{12} & 0 & 0 & 0 & 0 \\ \overline{c}_{21} & \overline{c}_{11} & 0 & 0 & 0 & 0 \\ 0 & 0 & 0 & 0 & 0 & 0 \\ 0 & 0 & 0 & c_{44} & 0 & 0 \\ 0 & 0 & 0 & 0 & c_{44} & 0 \\ 0 & 0 & 0 & 0 & 0 & c_{66} \end{pmatrix}, \tag{5.71}$$

$$[\overline{e}_{iq}] = \begin{bmatrix} 0 & 0 & 0 & 0 & e_{15} & 0 \\ 0 & 0 & 0 & e_{15} & 0 & 0 \\ \overline{e}_{31} & \overline{e}_{31} & 0 & 0 & 0 & 0 \end{bmatrix}, \quad [\overline{\varepsilon}_{kl}] = \begin{bmatrix} \varepsilon_{11} & 0 & 0 \\ 0 & \varepsilon_{11} & 0 \\ 0 & 0 & \overline{\varepsilon}_{33} \end{bmatrix}, \tag{5.72}$$

where

$$\begin{aligned} \overline{c}_{11} &= c_{11} - (c_{13})^2/c_{33}, \\ \overline{c}_{12} &= c_{12} - (c_{13})^2/c_{33}, \\ \overline{e}_{31} &= e_{31} - e_{33}c_{13}/c_{33}, \\ \overline{\varepsilon}_{33} &= \varepsilon_{33} + e_{33}^2/c_{33}. \end{aligned} \tag{5.73}$$

The stress relaxation in Eqs. (5.46) or (5.61) yields the following results:

$$[\gamma_{rs}] = \begin{bmatrix} \overline{c}_{11} & \overline{c}_{12} & 0 \\ \overline{c}_{12} & \overline{c}_{11} & 0 \\ 0 & 0 & c_{66} \end{bmatrix}, \quad [\psi_{ks}] = \begin{bmatrix} 0 & 0 & 0 \\ 0 & 0 & 0 \\ \overline{e}_{31} & \overline{e}_{31} & 0 \end{bmatrix},$$
$$[\zeta_{kj}] = \begin{bmatrix} \zeta_{11} & 0 & 0 \\ 0 & \zeta_{11} & 0 \\ 0 & 0 & \overline{\varepsilon}_{33} \end{bmatrix}, \quad r,s = 1,2,6, \quad k,j = 1,2,3, \tag{5.74}$$

where

$$\zeta_{11} = \varepsilon_{11} + e_{15}^2/c_{44}. \tag{5.75}$$

For plates made from crystals of class (6mm), the two thickness-shear correction factors are equal, i.e., $\kappa_1 = \kappa_2 = \kappa$. The value of κ will be determined in the next section.

The constitutive relations in Eqs. (5.58), (5.62), (5.38), and (5.40) take the following form for extension:

$$\begin{aligned} T_{11}^{(0)} &= 2h\left(\overline{c}_{11}u_{1,1}^{(0)} + \overline{c}_{12}u_{2,2}^{(0)} + \overline{e}_{31}\varphi^{(1)}\right), \\ T_{22}^{(0)} &= 2h\left(\overline{c}_{12}u_{1,1}^{(0)} + \overline{c}_{11}u_{2,2}^{(0)} + \overline{e}_{31}\varphi^{(1)}\right), \\ T_{12}^{(0)} &= 2hc_{66}\left(u_{1,2}^{(0)} + u_{2,1}^{(0)}\right), \end{aligned} \tag{5.76}$$

bending:

$$\begin{aligned} T_{3a}^{(0)} &= 2h\left[\kappa^2 c_{44}\left(u_{3,a}^{(0)} + u_a^{(1)}\right) + \kappa e_{15}\varphi_{,a}^{(0)}\right], \\ T_{11}^{(1)} &= \frac{2}{3}h^3\left(\overline{c}_{11}u_{1,1}^{(1)} + \overline{c}_{12}u_{2,2}^{(1)}\right), \\ T_{22}^{(1)} &= \frac{2}{3}h^3\left(\overline{c}_{12}u_{1,1}^{(1)} + \overline{c}_{11}u_{2,2}^{(1)}\right), \\ T_{12}^{(1)} &= \frac{2}{3}h^3 c_{66}\left(u_{1,2}^{(1)} + u_{2,1}^{(1)}\right), \end{aligned} \tag{5.77}$$

electrostatics:

$$\begin{aligned} D_a^{(0)} &= 2h\left[\kappa e_{15}\left(u_{3,a}^{(0)} + u_a^{(1)}\right) - \varepsilon_{11}\varphi_{,a}^{(0)}\right], \\ D_3^{(0)} &= 2h\left(\overline{e}_{31}u_{a,a}^{(0)} - \overline{\varepsilon}_{33}\varphi^{(1)}\right), \end{aligned} \tag{5.78}$$

$$D_a^{(1)} = -\frac{2}{3}h^3\zeta_{11}\varphi_{,a}^{(1)}, \tag{5.79}$$

and semiconduction:

$$\begin{aligned} J_a^{p(0)} &= 2hq\left(-p_0\mu_{11}^p\varphi_{,a}^{(0)} - d_{11}^p p_{,a}^{(0)}\right), \\ J_3^{p(0)} &= 2hq\left(-p_0\mu_{33}^p\varphi^{(1)} - d_{33}^p p^{(1)}\right), \end{aligned} \tag{5.80}$$

$$J_a^{p(1)} = \frac{2}{3}h^3 q\left(-p_0\mu_{11}^p\varphi_{,a}^{(1)} - d_{11}^p p_{,a}^{(1)}\right), \tag{5.81}$$

$$\begin{aligned} J_a^{n(0)} &= 2hq\left(-n_0\mu_{11}^n\varphi_{,a}^{(0)} + d_{11}^n n_{,a}^{(0)}\right), \\ J_3^{n(0)} &= 2hq\left(-n_0\mu_{33}^n\varphi^{(1)} + d_{33}^n n^{(1)}\right), \end{aligned} \tag{5.82}$$

$$J_a^{n(1)} = \frac{2}{3}h^3 q\left(-n_0\mu_{11}^n\varphi_{,a}^{(1)} + d_{11}^n n_{,a}^{(1)}\right), \tag{5.83}$$

where $\varphi^{(2)}$, $p^{(2)}$, and $n^{(2)}$ have been neglected.

The substitution of the above constitutive relations into Eqs. $(5.33)_1$, $(5.63)_2$, (5.34), (5.35), and (5.36) yields the following equations for extension described by $u_a^{(0)}$:

$$\begin{aligned} &\bar{c}_{11}u_{1,11}^{(0)} + c_{66}u_{1,22}^{(0)} + (\bar{c}_{12} + c_{66})u_{2,21}^{(0)} + \bar{e}_{31}\varphi_{,1}^{(1)} + \frac{1}{2h}t_1^{(0)} = \rho\ddot{u}_1^{(0)}, \\ &c_{66}u_{2,11}^{(0)} + \bar{c}_{11}u_{2,22}^{(0)} + (\bar{c}_{12} + c_{66})u_{1,12}^{(0)} + e_{15}\varphi_{,2}^{(1)} + \frac{1}{2h}t_2^{(0)} = \rho\ddot{u}_2^{(0)}, \end{aligned} \tag{5.84}$$

flexure described by $u_3^{(0)}$:

$$\kappa^2 c_{44}\left(u_{3,aa}^{(0)} + u_{a,a}^{(1)}\right) + \kappa e_{15}\varphi_{,aa}^{(0)} + \frac{1}{2h}t_3^{(0)} = \rho\ddot{u}_3^{(0)}, \tag{5.85}$$

thickness-shear described by $u_a^{(1)}$:

$$\begin{aligned} &\bar{c}_{11}u_{1,11}^{(1)} + c_{66}u_{1,22}^{(1)} + (\bar{c}_{12} + c_{66})u_{2,21}^{(1)} \\ &\quad -3h^{-2}\left[\kappa^2 c_{44}\left(u_1^{(1)} + u_{3,1}^{(0)}\right) + \kappa e_{15}\varphi_{,1}^{(0)}\right] + \frac{3}{2h^3}t_1^{(1)} = \rho\ddot{u}_1^{(1)}, \\ &c_{66}u_{2,11}^{(1)} + \bar{c}_{11}u_{2,22}^{(1)} + (\bar{c}_{12} + c_{66})u_{1,12}^{(1)} \\ &\quad -3h^{-2}\left[\kappa^2 c_{44}\left(u_2^{(1)} + u_{3,2}^{(0)}\right) + \kappa e_{15}\varphi_{,2}^{(0)}\right] + \frac{3}{2h^3}t_2^{(1)} = \rho\ddot{u}_2^{(1)}, \end{aligned} \tag{5.86}$$

electrostatics described by $\varphi^{(0)}$ and $\varphi^{(1)}$:

$$
\begin{aligned}
&-\varepsilon_{11}\varphi^{(0)}_{,aa} + \kappa e_{15}\left(u^{(0)}_{3,aa} + u^{(1)}_{a,a}\right) + \frac{1}{2h}d^{(0)} = q\left(p^{(0)} - n^{(0)}\right),\\
&-\zeta_{11}\varphi^{(1)}_{,aa} - 3h^{-2}\left(\overline{e}_{31}u^{(0)}_{a,a} - \overline{\varepsilon}_{33}\varphi^{(1)}\right) + \frac{3}{2h^3}d^{(1)} = q\left(p^{(1)} - n^{(1)}\right),
\end{aligned} \tag{5.87}
$$

and semiconduction described by $p^{(0)}$, $p^{(1)}$, $n^{(0)}$ and $n^{(1)}$:

$$
\begin{aligned}
&-p_0\mu^p_{11}\varphi^{(0)}_{,aa} - d^p_{11}p^{(0)}_{,aa} + \frac{1}{2hq}j^{p(0)} = -\dot{p}^{(0)},\\
&-p_0\mu^p_{11}\varphi^{(1)}_{,aa} - d^p_{11}p^{(1)}_{,aa}\\
&\quad -3h^{-2}\left(-p_0\mu^p_{33}\varphi^{(1)} - d^p_{33}p^{(1)}\right) + \frac{3}{2h^3q}j^{p(1)} = -\dot{p}^{(1)},
\end{aligned} \tag{5.88}
$$

$$
\begin{aligned}
&-n_0\mu^n_{11}\varphi^{(0)}_{,aa} + d^n_{11}n^{(0)}_{,aa} + \frac{1}{2hq}j^{n(0)} = \dot{n}^{(0)},\\
&-n_0\mu^n_{11}\varphi^{(1)}_{,aa} + d^n_{11}n^{(1)}_{,aa}\\
&\quad -3h^{-2}\left(-n_0\mu^n_{33}\varphi^{(1)} + d^n_{33}n^{(1)}\right) + \frac{3}{2h^3q}j^{n(1)} = \dot{n}^{(1)},
\end{aligned} \tag{5.89}
$$

with couplings among them.

5.8 Thickness-Shear Vibration

Consider thickness vibrations of the plate in Fig. 5.2. The plate is unelectroded at the top and bottom. There are no surface loads. The electric field in the surrounding free space is neglected.

Thickness vibrations are independent of x_1 and x_2. With the first-order plate equations we have, we can only study the fundamental thickness-shear vibration with one nodal plane (the x_1–x_2 plane) but not the higher-order thickness-shear vibrations with more nodal planes and thickness-stretch vibrations. We do not consider lateral field excitation so that the in-plane electric field is zero. For crystals of class (6mm), the fundamental thickness-shear motion is not coupled to the thickness electric field. In addition, crystals of class (6mm) are effectively

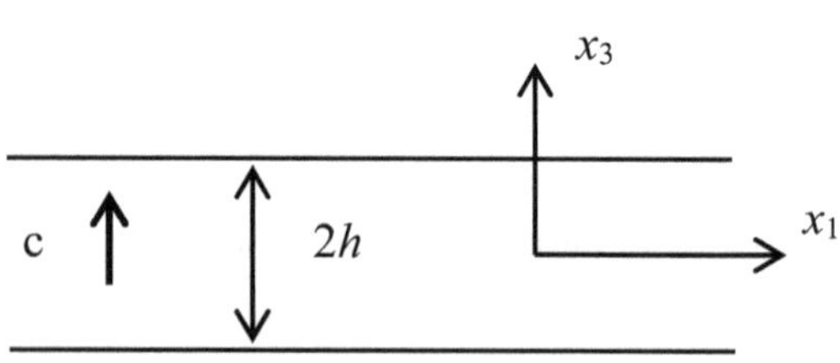

Fig. 5.2 An unbounded plate of crystals of class (6mm)

transversely isotropic in the x_1–x_2 plane. Therefore it is sufficient to study the u_1 displacement only. In this case, Eq. $(5.86)_1$ reduces to

$$-3h^{-2}\kappa^2 c_{44} u_1^{(1)} = \rho \ddot{u}_1^{(1)}. \tag{5.90}$$

For time-harmonic motions with a frequency ω, Eq. (5.90) determines

$$\omega^2 = \frac{3\kappa^2 c_{44}}{\rho h^2}. \tag{5.91}$$

The exact fundamental thickness-shear frequency determined from the three-dimensional equations is [6].

$$\omega^2 = \omega_\infty^2 = \frac{\pi^2 c_{44}}{4\rho h^2}. \tag{5.92}$$

Setting the frequencies in Eqs. (5.90) and (5.91) equal, we obtain the fundamental thickness-shear correction factor as

$$\kappa^2 = \frac{\pi^2}{12}. \tag{5.93}$$

5.9 Thickness-Shear Approximation

Thickness-shear vibration modes and standing as well as propagating waves are widely used in plate piezoelectric devices. These motions are described by $u_a^{(1)}$ and are usually accompanied by small flexural deformations described by $u_3^{(0)}$. The small flexural deformation can be eliminated by the so-called thickness-shear approximation [7] which simplifies the equations significantly. In this section we perform this approximation for unelectroded plates of crystals of class (6mm). We consider the case when there are no surface electromechanical loads. The electric field in the surrounding free space is neglected. We proceed as follows. From Eqs. (5.85) and $(5.87)_1$, we obtain, by eliminating $\varphi^{(0)}$,

$$\kappa^2\left(c_{44} + \frac{e_{15}^2}{\varepsilon_{11}}\right)\left(u_{3,aa}^{(0)} + u_{a,a}^{(1)}\right) - \frac{\kappa e_{15} q}{\varepsilon_{11}}\left(p^{(0)} - n^{(0)}\right) = \rho \ddot{u}_3^{(0)}. \tag{5.94}$$

Consider the following waves:

$$
\begin{aligned}
u_3^{(0)} &= A_3 \exp\left[i(\omega t + \xi_a x_a)\right], \\
u_b^{(1)} &= A_b \exp\left[i(\omega t + \xi_a x_a)\right], \\
p^{(0)} - n^{(0)} &= B \exp\left[i(\omega t + \xi_a x_a)\right],
\end{aligned}
\tag{5.95}
$$

where A_i and B are constants. The substitution of Eq. (5.95) into Eq. (5.94) results in

$$
\kappa^2 \left(c_{44} + \frac{e_{15}^2}{\varepsilon_{11}} \right)(-A_3 \xi_a \xi_a + A_a i \xi_a) - \frac{\kappa e_{15} q}{\varepsilon_{11}} B = -\rho \omega^2 A_3. \tag{5.96}
$$

We are interested in long waves with small wave numbers or small ξ_a. Then the term quadratic in ξ_a in the above equation can be neglected. For long thickness-shear waves, the frequency ω is very close to the exact fundamental thickness-shear frequency of an infinite plate given in Eq. (5.92). Hence we substitute Eq. (5.92) into Eq. (5.96). This leads to the following approximate version of Eq. (5.96):

$$
A_3 = -\frac{h^2}{3} \left(1 + \frac{e_{15}^2}{c_{44} \varepsilon_{11}} \right) i \xi_a A_a + \frac{4h^2 \kappa e_{15} q}{\pi^2 c_{44} \varepsilon_{11}} B, \tag{5.97}
$$

which is equivalent to the following differential relationship:

$$
u_3^{(0)} = -\frac{h^2}{3} (1 + k_{15}^2) u_{a,a}^{(1)} + \frac{4h^2 \kappa e_{15} q}{\pi^2 c_{44} \varepsilon_{11}} \left(p^{(0)} - n^{(0)} \right), \tag{5.98}
$$

where we have denoted the shear electromechanical coupling factor by

$$
k_{15}^2 = \frac{e_{15}^2}{c_{44} \varepsilon_{11}}. \tag{5.99}
$$

Substituting Eq. (5.98) into Eqs. (5.86) and $(5.87)_1$, neglecting the third- and higher-order derivatives of $u_3^{(0)}$ under the long-wave approximation, we obtain the following equations under the thickness-shear approximation:

$$
\begin{aligned}
&\widehat{c}_{11} u_{1,11}^{(1)} + c_{66} u_{1,22}^{(1)} + \widehat{c}_{12} u_{2,12}^{(1)} - \rho \omega_\infty^2 u_1^{(1)} \\
&\quad -3h^{-2} \kappa e_{15} \varphi_{,1}^{(0)} - \frac{\kappa e_{15} q}{\varepsilon_{11}} \left(p_{,1}^{(0)} - n_{,1}^{(0)} \right) = \rho \ddot{u}_1^{(1)}, \\
&c_{66} u_{2,11}^{(1)} + \widehat{c}_{11} u_{2,22}^{(1)} + \widehat{c}_{12} u_{1,12}^{(1)} - \rho \omega_\infty^2 u_2^{(1)} \\
&\quad -3h^{-2} \kappa e_{15} \varphi_{,2}^{(0)} - \frac{\kappa e_{15} q}{\varepsilon_{11}} \left(p_{,2}^{(0)} - n_{,2}^{(0)} \right) = \rho \ddot{u}_2^{(1)}, \\
&-\varepsilon_{11} \varphi_{,aa}^{(0)} + \kappa e_{15} u_{a,a}^{(1)} + k_{15}^2 \frac{h^2}{3} q \left(p_{,aa}^{(0)} - n_{,aa}^{(0)} \right) = q \left(p^{(0)} - n^{(0)} \right),
\end{aligned}
\tag{5.100}
$$

where we have denoted

$$\begin{aligned}\widehat{c}_{11} &= \overline{c}_{11} + \kappa^2 c_{44}\left(1 + k_{15}^2\right),\\ \widehat{c}_{12} &= \overline{c}_{12} + c_{66} + \kappa^2 c_{44}\left(1 + k_{15}^2\right).\end{aligned} \tag{5.101}$$

5.10 Propagation of Thickness-Shear Waves

We consider straight-crested waves propagating in the x_1 direction of a p-doped plate of crystals of class (6mm) with $u_2 = 0$ and $\partial/\partial x_2 = 0$ [1]. Then Eqs. (5.100) and $(5.88)_1$ reduce to

$$\begin{aligned}&\widehat{c}_{11} u_{1,11}^{(1)} - \rho\omega_\infty^2 u_1^{(1)} - 3h^{-2}\kappa e_{15}\varphi_{,1}^{(0)} - \frac{\kappa e_{15} q}{\varepsilon_{11}} p_{,1}^{(0)} = \rho \ddot{u}_1^{(1)},\\ &-\varepsilon_{11}\varphi_{,11}^{(0)} + \kappa e_{15} u_{1,1}^{(1)} + k_{15}^2 \frac{h^2}{3} q p_{,11}^{(0)} = q p^{(0)},\\ &\dot{p}^{(0)} - p_0 \mu_{11}^p \varphi_{,11}^{(0)} - d_{11}^p p_{,11}^{(0)} = 0.\end{aligned} \tag{5.102}$$

Let

$$\begin{aligned}u_1^{(1)} &= A\exp\left[i(\xi_1 x_1 - \omega t)\right],\\ \varphi^{(0)} &= B\exp\left[i(\xi_1 x_1 - \omega t)\right],\\ p^{(0)} &= C\exp\left[i(\xi_1 x_1 - \omega t)\right],\end{aligned} \tag{5.103}$$

where A, B, and C are undetermined constants. The substitution of Eq. (5.103) into Eq. (5.102) yields the following linear homogeneous equation for A, B, and C:

$$\begin{bmatrix}\rho\left(\omega^2 - \omega_\infty^2\right) - \widehat{c}_{11}\xi_1^2 & -3h^{-2}\kappa e_{15} i\xi_1 & -\dfrac{\kappa e_{15} q}{\varepsilon_{11}} i\xi_1\\ \kappa e_{15} i\xi_1 & \varepsilon_{11}\xi_1^2 & -k_{15}^2\dfrac{h^2}{3} q\xi_1^2 - q\\ 0 & p_0\mu_{11}^p\xi_1^2 & -i\omega + d_{11}^p\xi_1^2\end{bmatrix}\begin{Bmatrix}A\\ B\\ C\end{Bmatrix} = 0. \tag{5.104}$$

For nontrivial solutions the determinant of the coefficient matrix of Eq. (5.104) has to vanish, which gives the following equation that determines the dispersion relations of the wave:

$$\begin{aligned}\rho\omega^2 = {}& \rho\widehat{\omega}_\infty^2 + \widehat{c}_{11}\xi_1^2 - \frac{p_0 q \mu_{11}^p}{\varepsilon_{11}\left(d_{11}^p\xi_1^2 - i\omega\right)}\\ &\times\left\{\left(1 + \frac{1}{3}h^2 k_{15}^2\xi_1^2\right)\left[\rho\left(\omega^2 - \omega_\infty^2\right) - \xi_1^2\widehat{c}_{11}\right] + \kappa^2 k_{15}^2 c_{44}\xi_1^2\right\},\end{aligned} \tag{5.105}$$

where

$$\rho\widehat{\omega}_{\infty}^{2}=\rho\omega_{\infty}^{2}+\rho\frac{3\kappa^{2}c_{44}}{\rho h^{2}}\frac{e_{15}^{2}}{\varepsilon_{11}c_{44}}=\rho\omega_{\infty}^{2}\left(1+k_{15}^{2}\right) \tag{5.106}$$

is the piezoelectrically stiffened infinite plate thickness-shear frequency when there exists a coupling to the electric filed in the x_1 direction. When the semiconduction is small, Eq. (5.105) can be solved by an iteration or perturbation procedure. As the lowest or zero-order of approximation, we neglect the small semiconduction. Equation (5.105) reduces to the dispersion relation of long thickness-shear waves in a piezoelectric dielectric plate:

$$\rho\omega_{(0)}^{2}=\rho\widehat{\omega}_{\infty}^{2}+\widehat{c}_{11}\xi_{1}^{2}, \tag{5.107}$$

which is dispersive but not dissipative. For the next or first-order of approximation, we substitute Eq. (5.107) into the right-hand side of Eq. (5.105) and obtain

$$\begin{aligned}\rho\omega_{(1)}^{2}&=\rho\widehat{\omega}_{\infty}^{2}+\widehat{c}_{11}\xi_{1}^{2}-\frac{p_{0}q\mu_{11}^{p}}{\varepsilon_{11}\left(d_{11}^{p}\xi_{1}^{2}-i\omega_{(0)}\right)}\\&\times\left\{\left(1+\frac{1}{3}h^{2}k_{15}^{2}\xi_{1}^{2}\right)\left[\rho\left(\omega_{(0)}^{2}-\omega_{\infty}^{2}\right)-\xi_{1}^{2}\widehat{c}_{11}\right]+\kappa^{2}k_{15}^{2}c_{44}\xi_{1}^{2}\right\},\end{aligned} \tag{5.108}$$

which is complex and describes waves that are both dispersive and dissipative.

5.11 Equations for Shells

Shell structures are widely used in devices. The recent development of flexible electronics results in more devices with shell structures. In this section, we establish two-dimensional equations for piezoelectric semiconductor shells [8]. Consider a differential shell element (see Fig. 5.3). (α_1,α_2,α_3) is an orthogonal curvilinear coordinate system. α_1 and α_2 are the middle surface principal coordinates, and α_3 is the thickness coordinate. The thickness of the shell, $2h$, is much smaller than the principal radii of curvatures R_1 and R_2 of the middle surface. Let A_1 and A_2 be the Lamè coefficients corresponding to α_1 and α_2 at the middle surface. The metric in (α_1,α_2,α_3) is given by $A_1(1 + \alpha_3/R_1)$, $A_2(1 + \alpha_3/R_2)$, and 1. Equations (5.1), (5.2), (5.3), (5.4), (5.5), (5.6), (5.7), and (5.8) are still valid when they are considered as tensor equations and generalized into curvilinear coordinate systems. They are used to derive two-dimensional shell equations in this section.

The strain-displacement relations take the following form [9]:

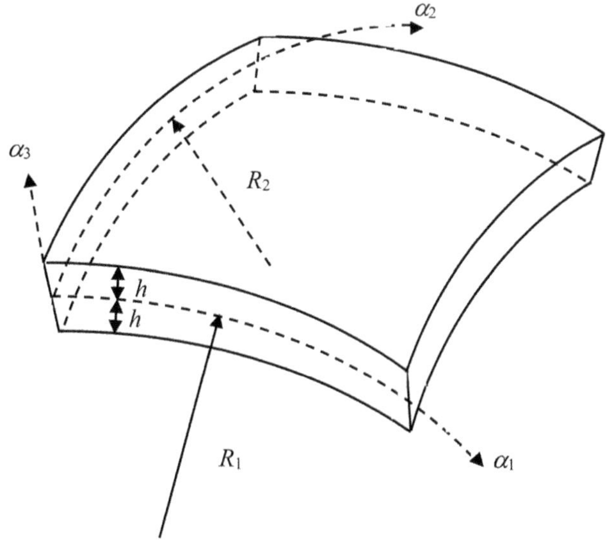

Fig. 5.3 A differential element of a thin shell and coordinate system

$$
\begin{aligned}
S_{11} &= \frac{1}{A_1\left(1+\dfrac{\alpha_3}{R_1}\right)}\left[\frac{\partial u_1}{\partial \alpha_1}+\frac{u_2}{A_2}\frac{\partial A_1}{\partial \alpha_2}+\frac{A_1 u_3}{R_1}\right],\\
S_{22} &= \frac{1}{A_2\left(1+\dfrac{\alpha_3}{R_2}\right)}\left[\frac{\partial u_2}{\partial \alpha_2}+\frac{u_1}{A_1}\frac{\partial A_2}{\partial \alpha_1}+\frac{A_2 u_3}{R_2}\right],\quad S_{33}=\frac{\partial u_3}{\partial \alpha_3},
\end{aligned}
\tag{5.109}
$$

$$
2S_{23} = A_2\left(1+\frac{\alpha_3}{R_2}\right)\frac{\partial}{\partial \alpha_3}\left[\frac{u_2}{A_2\left(1+\frac{\alpha_3}{R_2}\right)}\right]+\frac{1}{A_2\left(1+\frac{\alpha_3}{R_2}\right)}\frac{\partial u_3}{\partial \alpha_2}
\tag{5.110}
$$

$$
2S_{31} = A_1\left(1+\frac{\alpha_3}{R_1}\right)\frac{\partial}{\partial \alpha_3}\left[\frac{u_1}{A_1\left(1+\frac{\alpha_3}{R_1}\right)}\right]+\frac{1}{A_1\left(1+\frac{\alpha_3}{R_1}\right)}\frac{\partial u_3}{\partial \alpha_1}
\tag{5.111}
$$

$$
2S_{12} = \frac{A_1\left(1+\frac{\alpha_3}{R_1}\right)}{A_2\left(1+\frac{\alpha_3}{R_2}\right)}\frac{\partial}{\partial \alpha_2}\left[\frac{u_1}{A_1\left(1+\frac{\alpha_3}{R_1}\right)}\right]+\frac{A_2\left(1+\frac{\alpha_3}{R_2}\right)}{A_1\left(1+\frac{\alpha_3}{R_1}\right)}\frac{\partial}{\partial \alpha_1}\left[\frac{u_2}{A_2\left(1+\frac{\alpha_3}{R_2}\right)}\right].
\tag{5.112}
$$

The gradient of a scalar field, e.g., the electric potential, in the shell coordinates is given by

$$E_1 = -\frac{1}{A_1\left(1+\frac{\alpha_3}{R_1}\right)}\frac{\partial\varphi}{\partial\alpha_1},\quad E_2 = -\frac{1}{A_2\left(1+\frac{\alpha_3}{R_2}\right)}\frac{\partial\varphi}{\partial\alpha_2},\quad E_3 = -\frac{\partial\varphi}{\partial\alpha_3}. \tag{5.113}$$

The divergence of a vector field, e.g., the electric displacement vector, in the shell coordinates is

$$\begin{aligned}\nabla\cdot\mathbf{D} = &\frac{1}{A_1\left(1+\frac{\alpha_3}{R_1}\right)A_2\left(1+\frac{\alpha_3}{R_2}\right)}\left\{\frac{\partial}{\partial\alpha_1}\left[A_2\left(1+\frac{\alpha_3}{R_2}\right)D_1\right]\right.\\ &+\frac{\partial}{\partial\alpha_2}\left[A_1\left(1+\frac{\alpha_3}{R_1}\right)D_2\right]\\ &\left.+\frac{\partial}{\partial\alpha_3}\left[A_1\left(1+\frac{\alpha_3}{R_1}\right)A_2\left(1+\frac{\alpha_3}{R_2}\right)D_3\right]\right\}.\end{aligned} \tag{5.114}$$

For thin shells, if the dependence of $A_1(1 + \alpha_3/R_1)$ and $A_2(1 + \alpha_3/R_2)$ on α_3 is neglected after their derivatives in the above formulas with respect to α_3 have been carried out, Eqs. (5.109) through (5.114) reduce to

$$\begin{aligned}S_{11} &= \frac{1}{A_1}\left[\frac{\partial u_1}{\partial\alpha_1}+\frac{u_2}{A_2}\frac{\partial A_1}{\partial\alpha_2}+\frac{A_1u_3}{R_1}\right],\\ S_{22} &= \frac{1}{A_2}\left[\frac{\partial u_2}{\partial\alpha_2}+\frac{u_1}{A_1}\frac{\partial A_2}{\partial\alpha_1}+\frac{A_2u_3}{R_2}\right],\quad S_{33}=\frac{\partial u_3}{\partial\alpha_3},\end{aligned} \tag{5.115}$$

$$\begin{aligned}2S_{23} &= \frac{\partial u_2}{\partial\alpha_3}-\frac{u_2}{R_2}+\frac{1}{A_2}\frac{\partial u_3}{\partial\alpha_2},\\ 2S_{31} &= \frac{\partial u_1}{\partial\alpha_3}-\frac{u_1}{R_1}+\frac{1}{A_1}\frac{\partial u_3}{\partial\alpha_1},\\ 2S_{12} &= \frac{A_1}{A_2}\frac{\partial}{\partial\alpha_2}\left[\frac{u_1}{A_1}\right]+\frac{A_2}{A_1}\frac{\partial}{\partial\alpha_1}\left[\frac{u_2}{A_2}\right],\end{aligned} \tag{5.116}$$

$$E_1 = -\frac{1}{A_1}\frac{\partial\varphi}{\partial\alpha_1},\quad E_2 = -\frac{1}{A_2}\frac{\partial\varphi}{\partial\alpha_2},\quad E_3 = -\frac{\partial\varphi}{\partial\alpha_3}, \tag{5.117}$$

$$\nabla\cdot\mathbf{D} = \frac{1}{A_1A_2}\left\{\frac{\partial}{\partial\alpha_1}[A_2D_1]+\frac{\partial}{\partial\alpha_2}[A_1D_2]+\frac{\partial}{\partial\alpha_3}[A_1A_2D_3]+A_1A_2\left(\frac{1}{R_1}+\frac{1}{R_2}\right)D_3\right\}. \tag{5.118}$$

The power series expansion method for deriving plate equations can also be applied to shells. For a first-order shear deformation theory, we use the expansions in Eqs. (5.22) and (5.23) in which where $u_3^{(1)}$, $u_j^{(2)}$, $\varphi^{(2)}$, $p^{(2)}$, and $n^{(2)}$ will be eliminated through stress relaxation or simply neglected later. Under Eqs. (5.22) and (5.23), the strains, electric fields, and the gradients of carrier concentration perturbations for thin shells can be written as

$$S_{kl} = S_{kl}^{(0)} + \alpha_3 S_{kl}^{(1)}, \;\; E_k = E_k^{(0)} + \alpha_3 E_k^{(1)}, \tag{5.119}$$

$$P_k = P_k^{(0)} + \alpha_3 P_k^{(1)}, \;\; N_k = N_k^{(0)} + \alpha_3 N_k^{(1)}, \tag{5.120}$$

where

$$\begin{aligned} S_{11}^{(0)} &= \frac{1}{A_1}\left[\frac{\partial u_1^{(0)}}{\partial \alpha_1} + \frac{u_2^{(0)}}{A_2}\frac{\partial A_1}{\partial \alpha_2} + \frac{A_1 u_3^{(0)}}{R_1}\right], \\ S_{22}^{(0)} &= \frac{1}{A_2}\left[\frac{\partial u_2^{(0)}}{\partial \alpha_2} + \frac{u_1^{(0)}}{A_1}\frac{\partial A_2}{\partial \alpha_1} + \frac{A_2 u_3^{(0)}}{R_2}\right], \;\; S_{33}^{(0)} = u_3^{(1)}, \end{aligned} \tag{5.121}$$

$$\begin{aligned} 2S_{23}^{(0)} &= u_2^{(1)} - \frac{u_2^{(0)}}{R_2} + \frac{1}{A_2}\frac{\partial u_3^{(0)}}{\partial \alpha_2}, \\ 2S_{31}^{(0)} &= u_1^{(1)} - \frac{u_1^{(0)}}{R_1} + \frac{1}{A_1}\frac{\partial u_3^{(0)}}{\partial \alpha_1}, \\ 2S_{12}^{(0)} &= \frac{A_1}{A_2}\frac{\partial}{\partial \alpha_2}\left[\frac{u_1^{(0)}}{A_1}\right] + \frac{A_2}{A_1}\frac{\partial}{\partial \alpha_1}\left[\frac{u_2^{(0)}}{A_2}\right], \end{aligned} \tag{5.122}$$

$$\begin{aligned} S_{11}^{(1)} &= \frac{1}{A_1}\left[\frac{\partial u_1^{(1)}}{\partial \alpha_1} + \frac{u_2^{(1)}}{A_2}\frac{\partial A_1}{\partial \alpha_2} + \frac{A_1 u_3^{(1)}}{R_1}\right], \\ S_{22}^{(1)} &= \frac{1}{A_2}\left[\frac{\partial u_2^{(1)}}{\partial \alpha_2} + \frac{u_1^{(1)}}{A_1}\frac{\partial A_2}{\partial \alpha_1} + \frac{A_2 u_3^{(1)}}{R_2}\right], \\ S_{33}^{(1)} &= 2u_3^{(2)}, \end{aligned} \tag{5.123}$$

$$\begin{aligned} 2S_{23}^{(1)} &= 2u_2^{(2)} + \frac{1}{A_2}\frac{\partial u_3^{(1)}}{\partial \alpha_2} - \frac{u_2^{(1)}}{R_2}, \\ 2S_{31}^{(1)} &= 2u_1^{(2)} + \frac{1}{A_1}\frac{\partial u_3^{(1)}}{\partial \alpha_1} - \frac{u_1^{(1)}}{R_1}, \\ 2S_{12}^{(1)} &= \frac{A_1}{A_2}\frac{\partial}{\partial \alpha_2}\left[\frac{u_1^{(1)}}{A_1}\right] + \frac{A_2}{A_1}\frac{\partial}{\partial \alpha_1}\left[\frac{u_2^{(1)}}{A_2}\right], \end{aligned} \tag{5.124}$$

$$\begin{aligned} E_1^{(0)} &= -\frac{1}{A_1}\frac{\partial \varphi^{(0)}}{\partial \alpha_1}, \;\; E_2^{(0)} = -\frac{1}{A_2}\frac{\partial \varphi^{(0)}}{\partial \alpha_2}, \;\; E_3^{(0)} = -\varphi^{(1)}, \\ E_1^{(1)} &= -\frac{1}{A_1}\frac{\partial \varphi^{(1)}}{\partial \alpha_1}, \;\; E_2^{(1)} = -\frac{1}{A_2}\frac{\partial \varphi^{(1)}}{\partial \alpha_2}, \;\; E_3^{(1)} = -2\varphi^{(2)} \cong 0, \end{aligned} \tag{5.125}$$

$$
\begin{aligned}
P_1^{(0)} &= -\frac{1}{A_1}\frac{\partial p^{(0)}}{\partial \alpha_1}, \quad P_2^{(0)} = -\frac{1}{A_2}\frac{\partial p^{(0)}}{\partial \alpha_2}, \quad P_3^{(0)} = -p^{(1)}, \\
P_1^{(1)} &= -\frac{1}{A_1}\frac{\partial p^{(1)}}{\partial \alpha_1}, \quad P_2^{(1)} = -\frac{1}{A_2}\frac{\partial p^{(1)}}{\partial \alpha_2}, \quad P_3^{(1)} = -2p^{(2)} \cong 0,
\end{aligned}
\tag{5.126}
$$

$$
\begin{aligned}
N_1^{(0)} &= -\frac{1}{A_1}\frac{\partial n^{(0)}}{\partial \alpha_1}, \quad N_2^{(0)} = -\frac{1}{A_2}\frac{\partial n^{(0)}}{\partial \alpha_2}, \quad N_3^{(0)} = -n^{(1)}, \\
N_1^{(1)} &= -\frac{1}{A_1}\frac{\partial n^{(1)}}{\partial \alpha_1}, \quad N_2^{(1)} = -\frac{1}{A_2}\frac{\partial n^{(1)}}{\partial \alpha_2}, \quad N_3^{(1)} = -2n^{(2)} \cong 0.
\end{aligned}
\tag{5.127}
$$

$S_{ab}^{(0)}$ are the extensional and in-plane shear strains (in the tangent plane). $S_{a3}^{(0)}$ are the transverse shear strains related to bending. $S_{ab}^{(1)}$ are related to bending curvatures. The two-dimensional equations of motion are [9, 10]

$$
\begin{aligned}
&\frac{\partial(N_{11}A_2)}{\partial \alpha_1} + \frac{\partial(N_{21}A_1)}{\partial \alpha_2} + N_{12}\frac{\partial A_1}{\partial \alpha_2} - N_{22}\frac{\partial A_2}{\partial \alpha_1} + Q_{13}A_1A_2\frac{1}{R_1} + F_1^{(0)} \\
&= 2\rho h A_1 A_2 \ddot{u}_1^{(0)},
\end{aligned}
\tag{5.128}
$$

$$
\begin{aligned}
&\frac{\partial(N_{12}A_2)}{\partial \alpha_1} + \frac{\partial(N_{22}A_1)}{\partial \alpha_2} + N_{21}\frac{\partial A_2}{\partial \alpha_1} - N_{11}\frac{\partial A_1}{\partial \alpha_2} + Q_{23}A_1A_2\frac{1}{R_2} + F_2^{(0)} \\
&= 2\rho h A_1 A_2 \ddot{u}_2^{(0)},
\end{aligned}
\tag{5.129}
$$

$$
\begin{aligned}
&\frac{\partial}{\partial \alpha_1}(Q_{13}A_2) + \frac{\partial}{\partial \alpha_2}(Q_{23}A_1) - N_{11}\frac{A_1A_2}{R_1} - N_{22}\frac{A_1A_2}{R_2} + F_3^{(0)} \\
&\quad = 2\rho h A_1 A_2 \ddot{u}_3^{(0)},
\end{aligned}
\tag{5.130}
$$

$$
\begin{aligned}
&\frac{\partial(M_{11}A_2)}{\partial \alpha_1} + \frac{\partial(M_{21}A_1)}{\partial \alpha_2} + M_{12}\frac{\partial A_1}{\partial \alpha_2} - M_{22}\frac{\partial A_2}{\partial \alpha_1} - Q_{31}A_1A_2 + F_1^{(1)} \\
&= \frac{2\rho h^3}{3} A_1 A_2 \ddot{u}_1^{(1)},
\end{aligned}
\tag{5.131}
$$

$$
\begin{aligned}
&\frac{\partial(M_{12}A_2)}{\partial \alpha_1} + \frac{\partial(M_{22}A_1)}{\partial \alpha_2} + M_{21}\frac{\partial A_2}{\partial \alpha_1} - M_{11}\frac{\partial A_1}{\partial \alpha_2} - Q_{32}A_1A_2 + F_2^{(1)} \\
&= \frac{2\rho h^3}{3} A_1 A_2 \ddot{u}_2^{(1)}.
\end{aligned}
\tag{5.132}
$$

N_{ab} are the extensional and in-plane shear forces. $Q_{3c} = Q_{c3}$ are the transverse shear forces. M_{ab} are bending and twisting moments. Eqs. (5.128) and (5.129) are for extension, Eq. (5.130) for flexure, and Eqs. (5.131) and (5.132) for shear deformations in the α_1 and α_2 directions accompanying flexure. The two-dimensional charge equations are [6]

$$\frac{\partial\left(D_1^{(0)}A_2\right)}{\partial\alpha_1}+\frac{\partial\left(D_2^{(0)}A_1\right)}{\partial\alpha_2}+\left(\frac{1}{R_1}+\frac{1}{R_2}\right)A_1A_2D_3^{(0)}+D^{(0)}$$
$$=2hA_1A_2q\left(p^{(0)}-n^{(0)}\right), \tag{5.133}$$

$$\frac{\partial\left(D_1^{(1)}A_2\right)}{\partial\alpha_1}+\frac{\partial\left(D_2^{(1)}A_1\right)}{\partial\alpha_2}+\left(\frac{1}{R_1}+\frac{1}{R_2}\right)A_1A_2D_3^{(1)}-A_1A_2D_3^{(0)}+D^{(1)}$$
$$=\frac{2h^3}{3}qA_1A_2\left(p^{(1)}-n^{(1)}\right). \tag{5.134}$$

The two-dimensional continuity equations are:

$$\frac{\partial\left(J_1^{p(0)}A_2\right)}{\partial\alpha_1}+\frac{\partial\left(J_2^{p(0)}A_1\right)}{\partial\alpha_2}+\left(\frac{1}{R_1}+\frac{1}{R_2}\right)A_1A_2J_3^{p(0)}+J^{p(0)}$$
$$=-2hA_1A_2\dot{p}^{(0)}, \tag{5.135}$$

$$\frac{\partial\left(J_1^{p(1)}A_2\right)}{\partial\alpha_1}+\frac{\partial\left(J_2^{p(1)}A_1\right)}{\partial\alpha_2}+\left(\frac{1}{R_1}+\frac{1}{R_2}\right)A_1A_2J_3^{p(1)}-A_1A_2J_3^{p(0)}+J^{p(1)}$$
$$=-\frac{2h^3}{3}qA_1A_2\dot{p}^{(1)}, \tag{5.136}$$

$$\frac{\partial\left(J_1^{n(0)}A_2\right)}{\partial\alpha_1}+\frac{\partial\left(J_2^{n(0)}A_1\right)}{\partial\alpha_2}+\left(\frac{1}{R_1}+\frac{1}{R_2}\right)A_1A_2J_3^{n(0)}+J^{n(0)}$$
$$=2hA_1A_2\dot{n}^{(0)}, \tag{5.137}$$

$$\frac{\partial\left(J_1^{n(1)}A_2\right)}{\partial\alpha_1}+\frac{\partial\left(J_2^{n(1)}A_1\right)}{\partial\alpha_2}+\left(\frac{1}{R_1}+\frac{1}{R_2}\right)A_1A_2J_3^{n(1)}-A_1A_2J_3^{n(0)}+J^{n(1)}$$
$$=\frac{2h^3}{3}qA_1A_2\dot{n}^{(1)}. \tag{5.138}$$

The mechanical resultants and surface loads are defined by

$$\begin{aligned}
&\{N_{ab},Q_{3c},Q_{c3}\}=\left\{T_{ab}^{(0)},T_{3c}^{(0)},T_{c3}^{(0)}\right\}=\int_{-h}^{h}\{T_{ab},T_{3c},T_{c3}\}d\alpha_3,\\
&M_{ab}=T_{ab}^{(1)}=\int_{-h}^{h}T_{ab}\alpha_3d\alpha_3,\\
&F_j^{(n)}=\left[\alpha_3^nT_{3j}A_1A_2\right]_{-h}^{h},\ \ n=0,1,\ \ a,b,c=1,2.
\end{aligned} \tag{5.139}$$

The resultants of the electric displacement and surface charge are, for $m = 1$ and 2,

$$D_k^{(m)} = \int_{-h}^{h} D_k \alpha_3^m d\alpha_3, \quad D^{(m)} = \left[\alpha_3^m D_3 A_1 A_2\right]_{-h}^{h}. \tag{5.140}$$

The current resultants and surface currents are

$$\begin{aligned} J_k^{p(m)} &= \int_{-h}^{h} J_k^p \alpha_3^m d\alpha_3, \quad J^{p(m)} = \left[\alpha_3^m J_3^p A_1 A_2\right]_{-h}^{h}, \\ J_k^{n(m)} &= \int_{-h}^{h} J_k^n \alpha_3^m d\alpha_3, \quad J^{n(m)} = \left[\alpha_3^m J_3^n A_1 A_2\right]_{-h}^{h}. \end{aligned} \tag{5.141}$$

For shell constitutive relations, we substitute the three-dimensional constitutive relations in Eqs. (5.3) and (5.4) into Eqs. (5.139), (5.140), and (5.141), and perform the following stress relaxations:

$$\begin{aligned} T_{33}^{(0)} &= \int_{-h}^{h} T_{33} d\alpha_3 = 0, \\ T_{3j}^{(1)} &= \int_{-h}^{h} T_{33} \alpha_3 d\alpha_3 = 0, \quad j = 1, 2, 3, \end{aligned} \tag{5.142}$$

so that $S_{33}^{(0)}$ and $S_{3j}^{(1)}$, and hence $u_3^{(1)}$ and $u_j^{(2)}$ are eliminated in the same procedure as that in Sect. 5.5. Shear correction factors can be introduced in the manner of Eq. (5.57). Then the following two-dimensional constitutive relations are obtained (see Eqs. (5.58), (5.62), (5.38), and (5.40)):

$$\begin{aligned} T_{ij}^{(0)} &= 2h\left(c_{ijkl}^{(0)} S_{kl}^{(0)} - e_{kij}^{(0)} E_k^{(0)}\right), \\ D_k^{(0)} &= 2h\left(e_{kij}^{(0)} S_{ij}^{(0)} + \bar{\varepsilon}_{kl} E_l^{(0)}\right), \end{aligned} \tag{5.143}$$

$$\begin{aligned} T_r^{(1)} &= \frac{2h^3}{3}\left(\gamma_{rs} S_s^{(1)} - \psi_{kr} E_k^{(1)}\right), \\ D_i^{(1)} &= \frac{2h^3}{3}\left(\psi_{is} S_s^{(1)} + \zeta_{ij} E_j^{(1)}\right), \quad r, s = 1, 2, 6, \end{aligned} \tag{5.144}$$

$$\begin{aligned} J_i^{p(0)} &= 2h\left(\mu_{ij}^p E_j^{(0)} - D_{ij}^p P_j^{(0)}\right), \\ J_i^{n(0)} &= 2h\left(\mu_{ij}^n E_j^{(0)} + D_{ij}^n N_j^{(0)}\right), \end{aligned} \tag{5.145}$$

$$\begin{aligned} J_i^{p(1)} &= \frac{2h^3}{3}\left(\mu_{ij}^p E_j^{(1)} - D_{ij}^p P_j^{(1)}\right), \\ J_i^{n(1)} &= \frac{2h^3}{3}\left(\mu_{ij}^n E_j^{(1)} + D_{ij}^n N_j^{(1)}\right). \end{aligned} \tag{5.146}$$

The effective shell material constants after stress relaxations are defined in Eqs. (5.59), (5.60), and (5.50). In summary, with successive substitutions, Eqs. (5.128), (5.129), (5.130), (5.131), (5.132), (5.133), (5.134), (5.135), (5.136), (5.137), and (5.138) can be written as 11 equations for $u_i^{(0)}$, $u_a^{(1)}$, $\varphi^{(0)}$, $\varphi^{(1)}$, $p^{(0)}$, $p^{(1)}$, $n^{(0)}$, and $n^{(1)}$. At the boundary of a shell with an in-plane unit exterior normal **n** and an in-plane unit tangent **s**, we may prescribe

$$\begin{aligned} &N_{nn} \text{ or } u_n^{(0)}, \ N_{ns} \text{ or } u_s^{(0)}, \ Q_{n3}^{(0)} \text{ or } u_3^{(0)}, \\ &M_{nn} \text{ or } u_n^{(1)}, \ M_{ns} \text{ or } u_s^{(1)}, \\ &D_n^{(0)} \text{ or } \varphi^{(0)}, \ D_n^{(1)} \text{ or } \varphi^{(1)}, \end{aligned} \tag{5.147}$$

$$\begin{aligned} &J_n^{p(0)} \text{ or } p^{(0)}, \ J_n^{p(1)} \text{ or } p^{(1)}, \\ &J_n^{n(0)} \text{ or } n^{(0)}, \ J_n^{n(1)} \text{ or } n^{(1)}. \end{aligned} \tag{5.148}$$

For the classical theory of a shell in coupled extension and bending without shear deformations, we set the relevant shear strains to zero:

$$\begin{aligned} 2S_{31}^{(0)} &= u_1^{(1)} - \frac{u_1^{(0)}}{R_1} + \frac{1}{A_1}\frac{\partial u_3^{(0)}}{\partial \alpha_1} = 0, \\ 2S_{23}^{(0)} &= u_2^{(1)} - \frac{u_2^{(0)}}{R_2} + \frac{1}{A_2}\frac{\partial u_3^{(0)}}{\partial \alpha_2} = 0. \end{aligned} \tag{5.149}$$

This allows us to express $u_1^{(1)}$ and $u_2^{(1)}$ in terms of the zero-order displacements for extension and bending. Furthermore, we ignore the rotatory inertia $2\rho h^3/3$ in Eqs. (5.131) and (5.132) to obtain

$$\frac{\partial(M_{11}A_2)}{\partial \alpha_1} + \frac{\partial(M_{21}A_1)}{\partial \alpha_2} + M_{12}\frac{\partial A_1}{\partial \alpha_2} - M_{22}\frac{\partial A_2}{\partial \alpha_1} - Q_{31}A_1A_2 + F_1^{(1)} = 0, \tag{5.150}$$

$$\frac{\partial(M_{12}A_2)}{\partial \alpha_1} + \frac{\partial(M_{22}A_1)}{\partial \alpha_2} + M_{21}\frac{\partial A_2}{\partial \alpha_1} - M_{11}\frac{\partial A_1}{\partial \alpha_2} - Q_{32}A_1A_2 + F_2^{(1)} = 0, \tag{5.151}$$

which yields expressions for the transverse shear forces Q_{31} and Q_{32} in terms of the bending and twisting moments, or effectively the constitutive relations for Q_{31} and Q_{32}. The mechanical equations of motion for the classical theory are left to be Eqs. (5.128), (5.129), and (5.130) for $u_i^{(0)}$:

$$\begin{aligned} &\frac{\partial(N_{11}A_2)}{\partial \alpha_1} + \frac{\partial(N_{21}A_1)}{\partial \alpha_2} + N_{12}\frac{\partial A_1}{\partial \alpha_2} - N_{22}\frac{\partial A_2}{\partial \alpha_1} + Q_{13}A_1A_2\frac{1}{R_1} + F_1^{(0)} \\ &= 2\rho h A_1 A_2 \ddot{u}_1^{(0)}, \end{aligned} \tag{5.152}$$

$$\frac{\partial(N_{12}A_2)}{\partial\alpha_1}+\frac{\partial(N_{22}A_1)}{\partial\alpha_2}+N_{21}\frac{\partial A_2}{\partial\alpha_1}-N_{11}\frac{\partial A_1}{\partial\alpha_2}+Q_{23}A_1A_2\frac{1}{R_2}+F_2^{(0)}$$
$$=2\rho hA_1A_2\ddot{u}_2^{(0)}, \tag{5.153}$$

$$\frac{\partial}{\partial\alpha_1}(Q_{13}A_2)+\frac{\partial}{\partial\alpha_2}(Q_{23}A_1)-N_{11}\frac{A_1A_2}{R_1}-N_{22}\frac{A_1A_2}{R_2}+F_3^{(0)}$$
$$=2\rho hA_1A_2\ddot{u}_3^{(0)}. \tag{5.154}$$

A thin shell can sometimes be treated as a membrane that does not resist bending, without bending moments and transverse shear forces. In this case Eqs. (5.152), (5.123), and (5.154) reduce the following membrane theory:

$$\frac{\partial(N_{11}A_2)}{\partial\alpha_1}+\frac{\partial(N_{21}A_1)}{\partial\alpha_2}+N_{12}\frac{\partial A_1}{\partial\alpha_2}-N_{22}\frac{\partial A_2}{\partial\alpha_1}+F_1^{(0)}=2\rho hA_1A_2\ddot{u}_1^{(0)}, \tag{5.155}$$

$$\frac{\partial(N_{12}A_2)}{\partial\alpha_1}+\frac{\partial(N_{22}A_1)}{\partial\alpha_2}+N_{21}\frac{\partial A_2}{\partial\alpha_1}-N_{11}\frac{\partial A_1}{\partial\alpha_2}+F_2^{(0)}=2\rho hA_1A_2\ddot{u}_2^{(0)}, \tag{5.156}$$

$$-N_{11}\frac{1}{R_1}-N_{22}\frac{1}{R_2}+F_3^{(0)}=2\rho h\ddot{u}_3^{(0)}. \tag{5.157}$$

References

1. J.S. Yang, H.G. Zhou, Amplification of acoustic waves in piezoelectric semiconductor plates. Int. J. Solids Struct. **42**, 3171–3183 (2005)
2. Y.X. Luo, C.L. Zhang, W.Q. Chen, J.S. Yang, Piezotronic effect of a thin film with elastic and piezoelectric semiconductor layers under a static flexural loading. ASME J. Appl. Mech. **86**, 051003 (2019)
3. R.D. Mindlin, in *An Introduction to the Mathematical Theory of Vibrations of Elastic Plates*, ed. by J.S. Yang, (World Scientific, Singapore, 2006)
4. R.D. Mindlin, High frequency vibrations of crystal plates. Q. Appl. Math. **19**, 51–61 (1961)
5. R.D. Mindlin, High frequency vibrations of piezoelectric crystal plates. Int. J. Solids Struct. **8**, 895–906 (1972)
6. J.S. Yang, *The Mechanics of Piezoelectric Structures* (World Scientific, Singapore, 2006)
7. J.L. Bleustein, H.F. Tiersten, Forced thickness-shear vibrations of discontinuously plated piezoelectric plates. J. Acoust. Soc. Am. **43**, 1311–1318 (1968)
8. J.S. Yang, X.M. Yang, J.A. Turner, Amplification of acoustic waves in piezoelectric semiconductor shells. J. Intell. Mater. Syst. Struct. **16**, 613–621 (2005)
9. H. Kraus, *Thin Elastic Shells* (Wiley, New York, 1967)
10. H.S. Tzou, *Piezoelectric Shells* (Kluwer, Dordrecht, 1993)

Chapter 6
Composite Structures

In acoustoelectric and piezotronic devices, the redistribution or motion of charge carriers is caused or affected by mechanical loads through the electric field produced via piezoelectric coupling. This coupling can also happen in composite structures of piezoelectric dielectrics and nonpiezoelectric semiconductors. In this chapter we study a few problems of this type.

6.1 Surface Waves

Consider a p-doped nonpiezoelectric semiconductor plate of thickness $2h$ on a piezoelectric dielectric half space as shown in Fig. 6.1 along with the coordinate system. The plate is unelectroded and is without free charges or currents at its surfaces. We study surface waves propagating in the x_1 direction [1].

Consider the piezoelectric half space first. We limit ourselves to the case of ceramics poled in the x_3 direction. These materials allow the following anti-plane motions described by

$$u_1 = u_2 = 0, \quad u_3 = u_3(x_1, x_2, t), \quad \varphi = \varphi(x_1, x_2, t), \tag{6.1}$$

$$\begin{Bmatrix} 2S_{13} \\ 2S_{23} \end{Bmatrix} = \nabla u_3, \quad \begin{Bmatrix} E_1 \\ E_2 \end{Bmatrix} = -\nabla \varphi, \tag{6.2}$$

$$\begin{Bmatrix} T_{13} \\ T_{23} \end{Bmatrix} = c_{44} \nabla u_3 + e_{15} \nabla \varphi, \quad \begin{Bmatrix} D_1 \\ D_2 \end{Bmatrix} = e_{15} \nabla u_3 - \varepsilon_{11} \nabla \varphi, \tag{6.3}$$

$$\begin{aligned} c_{44} \nabla^2 u_3 + e_{15} \nabla^2 \varphi &= \rho \ddot{u}_3, \\ e_{15} \nabla^2 u_3 - \varepsilon_{11} \nabla^2 \varphi &= 0, \end{aligned} \tag{6.4}$$

J. Yang, *Analysis of Piezoelectric Semiconductor Structures*,
https://doi.org/10.1007/978-3-030-48206-0_6

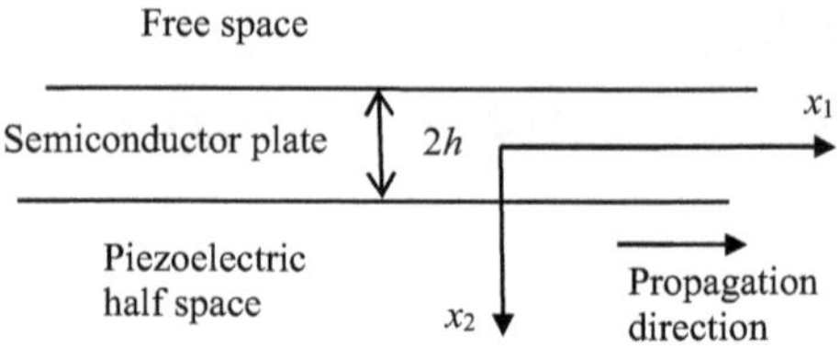

Fig. 6.1 A piezoelectric dielectric half space with a nonpiezoelectric semiconductor plate

where $\nabla = \mathbf{i}_1\partial_1 + \mathbf{i}_2\partial_2$ is the two-dimensional gradient operator and $\nabla^2 = \partial_1^2 + \partial_2^2$ is the two-dimensional Laplacian. Introduce [2]

$$\psi = \varphi - \frac{e_{15}}{\varepsilon_{11}} u_3, \tag{6.5}$$

then

$$\begin{aligned} T_{23} &= \bar{c}_{44} u_{3,2} + e_{15}\psi_{,2}, \\ T_{31} &= \bar{c}_{44} u_{3,1} + e_{15}\psi_{,1}, \end{aligned} \tag{6.6}$$

$$\begin{aligned} D_1 &= -\varepsilon_{11}\psi_{,1}, \\ D_2 &= -\varepsilon_{11}\psi_{,2}, \end{aligned} \tag{6.7}$$

$$\begin{aligned} &\bar{c}_{44}\nabla^2 u_3 = \rho \ddot{u}_3, \\ &\nabla^2 \psi = 0, \end{aligned} \tag{6.8}$$

where

$$\bar{c}_{44} = c_{44} + \frac{e_{15}^2}{\varepsilon_{11}} = c_{44}\left(1 + k_{15}^2\right), \quad k_{15}^2 = \frac{e_{15}^2}{\varepsilon_{11} c_{44}}. \tag{6.9}$$

For a surface wave solution, we require that

$$u_3, \ \varphi \to 0, \quad x_2 \to +\infty. \tag{6.10}$$

Consider possible solutions in the following form:

$$\begin{aligned} u_3 &= A \exp\left[-\xi_2(x_2 - h)\right] \exp\left[i(\xi_1 x_1 - \omega t)\right], \\ \psi &= B \exp\left[-\xi_1(x_2 - h)\right] \exp\left[i(\xi_1 x_1 - \omega t)\right], \end{aligned} \tag{6.11}$$

where A and B are undetermined constants and ξ_2 should be positive for decaying behavior away from the surface. Equation $(6.11)_2$ already satisfies Eq. $(6.8)_2$. For Eq. $(6.11)_1$ to satisfy Eq. $(6.8)_1$, we must have

$$\bar{c}_{44}\left(\xi_1^2 - \xi_2^2\right) = \rho\omega^2, \tag{6.12}$$

which leads to the following expression for ξ_2:

$$\xi_2^2 = \xi_1^2 - \frac{\rho\omega^2}{\overline{c}_{44}} = \xi_1^2\left(1 - \frac{v^2}{v_T^2}\right) > 0, \tag{6.13}$$

where

$$v^2 = \frac{\omega^2}{\xi_1^2}, \quad v_T^2 = \frac{\overline{c}_{44}}{\rho}. \tag{6.14}$$

The following fields are needed in the boundary and continuity conditions:

$$\begin{aligned} \varphi &= \left\{ B\exp\left[-\xi_1(x_2 - h)\right] + \frac{e_{15}}{\varepsilon_{11}} A\exp\left[-\xi_2(x_2 - h)\right] \right\} \\ &\quad \times \exp\left[i(\xi_1 x_1 - \omega t)\right], \\ T_{23} &= -\{A\overline{c}_{44}\xi_2 \exp\left[-\xi_2(x_2 - h)\right] + e_{15}B\xi_1 \exp\left[-\xi_1(x_2 - h)\right]\} \\ &\quad \times \exp\left[i(\xi_1 x_1 - \omega t)\right], \\ D_2 &= \varepsilon_{11}B\xi_1 \exp\left[-\xi_1(x_2 - h)\right] \exp\left[i(\xi_1 x_1 - \omega t)\right]. \end{aligned} \tag{6.15}$$

Electric fields can also exist in the free space of $x_2 < -h$, which are governed by

$$\begin{aligned} &\nabla^2\varphi = 0, \quad x_2 < -h, \\ &\varphi \to 0, \quad x_2 \to -\infty. \end{aligned} \tag{6.16}$$

A surface wave solution to Eq. (6.16) is

$$\varphi = C\exp\left[\xi_1(x_2 + h)\right] \exp\left[i(\xi_1 x_1 - \omega t)\right], \tag{6.17}$$

where C is an undetermined constant. From Eq. (6.17), in the free space, we have

$$D_2 = -\varepsilon_0\xi_1 C\exp\left[\xi_1(x_2 + h)\right] \exp\left[i(\xi_1 x_1 - \omega t)\right]. \tag{6.18}$$

For the semiconductor plate, since the normal of the plate is along x_2, the plate equations in Chap. 5 cannot be used directly. Fortunately we only need the zero-order equations of the plate which will be quickly derived in the following in the coordinate system in Fig. 6.1. The zero-order equations govern extensional motions described by $u_1(x_1, x_3, t)$ and $u_3(x_1, x_3, t)$. The dominating stress components are T_{11}, T_{33}, and T_{13}. Therefore we make the following stress relaxation for thin plates:

$$T_{2j} = 0, \quad j = 1, 2, 3. \tag{6.19}$$

According to the compact matrix notation, with the range of p, q as 1, 2, ... and 6, Eq. (6.19) can be written as

$$T_q = 0, \quad q = 2, 4, 6. \tag{6.20}$$

For convenience we introduce an index convention that subscripts u, v, and w take the values 2, 4, and 6 while subscripts r, s, and t take the remaining values 1, 3, and 5. Then, the constitutive relations for a nonpiezoelectric semiconductor plate can be written as

$$\begin{aligned} T_r &= c'_{rs} S_s + c'_{ru} S_u, \\ T_v &= c'_{vs} S_s + c'_{vw} S_w = 0, \\ D_i &= \varepsilon'_{ij} E_j \ , \end{aligned} \tag{6.21}$$

where we have used a prime to indicate the elastic and dielectric constants of the plate. From Eq. $(6.21)_2$, we obtain

$$S_u = -c'^{-1}_{uv} c'_{vs} S_s. \tag{6.22}$$

The substitution of Eq. (6.22) into Eq. $(6.21)_1$ gives the constitutive relations needed:

$$\begin{aligned} T_r &= \overline{c}'_{rs} S_s, \\ D_i &= \varepsilon'_{ij} E_j, \end{aligned} \tag{6.23}$$

where we have kept Eq. $(6.21)_3$ for convenience. In Eq. $(6.23)_1$, the thin-plate effective elastic constants are

$$\overline{c}'_{rs} = c'_{rs} - c'_{rv} c'^{-1}_{vw} c'_{ws}. \tag{6.24}$$

With subscripts a, b, c, and d assuming 1 and 3 but not 2, Eq. (6.23) can be written as

$$\begin{aligned} T_{ab} &= \overline{c}'_{abcd} S_{cd}, \\ D_i &= \varepsilon'_{ij} E_j. \end{aligned} \tag{6.25}$$

Integrating Eq. $(1.1)_1$ for $i = 1, 3$ when $\mathbf{f} = 0$ with respect to x_2 through the plate thickness, and also integrating Eq. $(1.10)_{1,2}$ along the plate thickness, we obtain the following equations for the plate:

$$
\begin{aligned}
&T_{ab,a}+\frac{1}{2h}\left[T_{2b}(x_2=h)-T_{2b}(x_2=-h)\right]=\rho\ddot{u}_b,\\
&D_{a,a}+\frac{1}{2h}\left[D_2(x_2=h)-D_2(x_2=-h)\right]=q(\Delta p),\\
&q\frac{\partial(\Delta p)}{\partial t}+J_{a,a}+\frac{1}{2h}\left[J_2(x_2=h)-J_2(x_2=-h)\right]=0,
\end{aligned}
\tag{6.26}
$$

where u_a, T_{ab}, D_a, J_a, and Δp are independent of x_2 and are understood to be the averages of the corresponding fields along the plate thickness. The specific motion of the plate relevant to the present problem is described by $u_3(x_1,t)$ and is called the face-shear (FS) wave of a plate. Consider the following wave solution:

$$
\begin{aligned}
&u_3=A\exp\left[i(\xi_1x_1-\omega t)\right],\quad \varphi=C\exp\left[i(\xi_1x_1-\omega t)\right],\\
&\Delta p=P\exp\left[i(\xi_1x_1-\omega t)\right],
\end{aligned}
\tag{6.27}
$$

where P is an undetermined constant. Equation (6.27) already satisfies the continuity of displacement between the plate and the ceramic half space and the continuity of electric potential between the plate and the free space. We consider a plate of silicon which is a cubic crystal of class (m3m). The elastic and dielectric constants are given by

$$
\begin{pmatrix}
c'_{11} & c'_{12} & c'_{12} & 0 & 0 & 0\\
c'_{12} & c'_{11} & c_{12} & 0 & 0 & 0\\
c'_{12} & c'_{12} & c'_{11} & 0 & 0 & 0\\
0 & 0 & 0 & c'_{44} & 0 & 0\\
0 & 0 & 0 & 0 & c'_{44} & 0\\
0 & 0 & 0 & 0 & 0 & c'_{44}
\end{pmatrix},
\begin{pmatrix}
\varepsilon'_{11} & 0 & 0\\
0 & \varepsilon'_{11} & 0\\
0 & 0 & \varepsilon'_{11}
\end{pmatrix}.
\tag{6.28}
$$

Then the constitutive relations take the following specific form:

$$
\begin{aligned}
&T_{13}=\overline{c}'_{55}S_{13}=\overline{c}'_{44}u_{3,1}=\overline{c}'_{44}i\xi_1A\exp\left[i(\xi_1x_1-\omega t)\right],\\
&D_1=\varepsilon'_{11}E_1=-\varepsilon'_{11}\varphi_{,1}=-\varepsilon'_{11}i\xi_1C\exp\left[i(\xi_1x_1-\omega t)\right],\\
&J_1^p=-qp_0\mu_{11}^p\varphi_{,1}-qD_{11}^p(\Delta p)_{,1}\\
&\quad=\left(-qp_0\mu_{11}^pi\xi_1C-qD_{11}^pi\xi_1P\right)\exp\left[i(\xi_1x_1-\omega t)\right].
\end{aligned}
\tag{6.29}
$$

The substitution of Eqs. (6.11), (6.15), (6.17), (6.18), (6.27), and (6.29) into the continuity condition of the electric potential between the ceramic half space and the plate, Eq. $(6.26)_1$ for $b=3$, and Eq. $(6.26)_{2,3}$ yields

$$\begin{aligned}
&B+\frac{e_{15}}{\varepsilon_{11}}A=C,\\
&-\bar{c}'_{44}\xi_1^2A-\frac{1}{2h}(A\bar{c}_{44}\xi_2+e_{15}B\xi_1)=-\rho'\omega^2A,\\
&\varepsilon'_{11}\xi_1^2C+\frac{1}{2h}(\varepsilon_{11}B\xi_1+\varepsilon_0\xi_1C)=qP,\\
&-qi\omega P+i\xi_1\left(-qp_0\mu^p_{11}i\xi_1C-qD^p_{11}i\xi_1P\right)=0,
\end{aligned}\tag{6.30}$$

which is a system of linear homogeneous equations for A, B, C, and P. For nontrivial solutions the determinant of the coefficient matrix has to vanish:

$$\begin{vmatrix}
\dfrac{e_{15}}{\varepsilon_{11}} & 1 & -1 & 0\\
\rho'\omega^2-\bar{c}'_{44}\xi_1^2-\dfrac{\bar{c}_{44}\xi_2}{2h} & -\dfrac{e_{15}\xi_1}{2h} & 0 & 0\\
0 & \dfrac{\varepsilon_{11}\xi_1}{2h} & \dfrac{\varepsilon_0\xi_1}{2h}+\varepsilon'_{11}\xi_1^2 & -q\\
0 & 0 & qp_0\mu^p_{11}\xi_1^2 & -qi\omega+qD^p_{11}\xi_1^2
\end{vmatrix}=0,\tag{6.31}$$

which determines the dispersion relations between ω and ξ_1 of the surface waves. In terms of the surface wave speed $v=\omega/\xi_1$, Eq. (6.31) can be written in the following form:

$$\begin{aligned}
&\left(\frac{v^2}{v'^2_T}-1\right)\frac{\bar{c}'_{44}}{\bar{c}_{44}}2h\xi_1-\sqrt{1-\frac{v^2}{v_T^2}}+\bar{k}_{15}^2\\
&=\frac{\bar{k}_{15}^2}{1+\dfrac{\varepsilon_0}{\varepsilon_{11}}+\dfrac{\varepsilon'_{11}}{\varepsilon_{11}}\xi_1 2h+\dfrac{qp_0\mu^p_{11}2h}{\varepsilon_{11}\left(D^p_{11}\xi_1-iv\right)}},
\end{aligned}\tag{6.32}$$

where Eq. (6.13) has been used, and

$$v'^2_T=\frac{\bar{c}'_{44}}{\rho'},\quad \bar{k}_{15}^2=\frac{e_{15}^2}{\varepsilon_{11}\bar{c}_{44}}.\tag{6.33}$$

We make the following observations from Eq. (6.32):

(i) When $h=0$, i.e., the semiconductor plate does not exist, Eq. (6.32) reduces to

$$v^2=v_T^2\left[1-\frac{\bar{k}_{15}^4}{\left(1+\varepsilon_{11}/\varepsilon_0\right)^2}\right]=v_{BG}^2,\tag{6.34}$$

which is the speed of the well-known Bleustein-Gulyaev wave [2], an anti-plane or shear-horizontal (SH) surface wave that relies on piezoelectric coupling and does not have an elastic counterpart.

(ii) When $\overline{k}_{15}^2 = 0$, i.e., the half space is non-piezoelectric, electromechanical coupling disappears, and the wave is purely elastic. In this case Eq. (6.32) reduces to

$$\left(\frac{v^2}{v_T'^2} - 1\right)\frac{\overline{c}_{44}'}{\overline{c}_{44}}2h\xi_1 - \sqrt{1 - \frac{v^2}{v_T^2}} = 0, \tag{6.35}$$

which is the equation that determines the speed of Love wave, an anti-plane surface wave in an elastic half space carrying an elastic plate, in the limit when the plate is very thin compared to the wavelength ($\xi_1 h < <1$). Love waves are known to exist when the relevant elastic stiffness of the plate is smaller than that of the half space.

(iii) When semiconduction is small, Eq. (6.32) can be solved by an iteration or perturbation procedure. As the lowest (zero-) order of approximation, we neglect the small semiconduction and denote the zero-order solution by $v_{(0)}$. Then, from Eq. (6.32),

$$\left(\frac{v_{(0)}^2}{v_T'^2} - 1\right)\frac{\overline{c}_{44}'}{\overline{c}_{44}}2h\xi_1 - \sqrt{1 - \frac{v_{(0)}^2}{v_T^2}} + \overline{k}_{15}^2 = \frac{\overline{k}_{15}^2}{1 + \frac{\varepsilon_0}{\varepsilon_{11}} + \frac{\varepsilon_{11}'}{\varepsilon_{11}}\xi_1 2h}, \tag{6.36}$$

which determines a dispersive wave. For the next order of approximation we substitute $v_{(0)}$ into the right-hand side of Eq. (6.32) and obtain the following equation for $v_{(1)}$:

$$\left(\frac{v_{(1)}^2}{v_T'^2} - 1\right)\frac{\overline{c}_{44}'}{\overline{c}_{44}}2h\xi_1 - \sqrt{1 - \frac{v_{(1)}^2}{v_T^2}} + \overline{k}_{15}^2 = \frac{\overline{k}_{15}^2}{1 + \frac{\varepsilon_0}{\varepsilon_{11}} + \frac{\varepsilon_{11}'}{\varepsilon_{11}}\xi_1 2h + \frac{qp_0\mu_{11}^p 2h}{\varepsilon_{11}\left(D_{11}^p\xi_1 - iv_{(0)}\right)}}, \tag{6.37}$$

which is complex and suggests a wave that is both dispersive and dissipative.

6.2 Extension of Rods

Consider the composite rod shown in Fig. 6.2. (x,y,z) correspond to (x_1,x_2,x_3). It consists of two piezoelectric dielectric layers such as polarized ceramics (PZT) and a nonpiezoelectric semiconductor layer such as silicon (Si). The composite rod is within $|z| < L$. It is under the action of a pair of equal and opposite axial forces F [3].

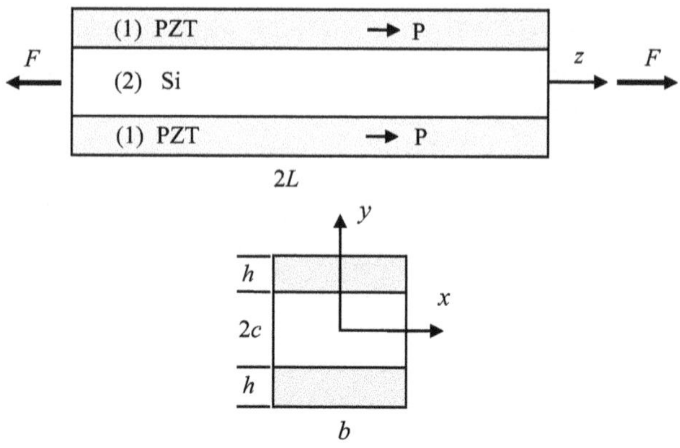

Fig. 6.2 Side view and cross section of a composite rod of piezoelectric dielectrics and nonpiezoelectric semiconductors

The rod is in extension along z. We denote the relevant axial fields by

$$\begin{aligned} &u_3 = u, \quad S_{33} = S, \quad T_{33} = T, \\ &E_3 = E, \quad D_3 = D, \\ &J_3^p = J^p, \quad J_3^n = J^n. \end{aligned} \tag{6.38}$$

For extension of thin rods we have, approximately,

$$u \cong u(z,t), \quad \varphi \cong \varphi(z,t). \tag{6.39}$$

The relevant strain-displacement relation and electric field-potential relation are

$$S = \frac{\partial u}{\partial z}, \quad E = -\frac{\partial \varphi}{\partial z}. \tag{6.40}$$

For polarized ceramics, the relevant constitutive relations are

$$\begin{aligned} S &= s_{33}^E T + d_{33} E, \\ D &= d_{33} T + \varepsilon_{33}^T E, \end{aligned} \tag{6.41}$$

where $T_1 = T_2 = 0$ has been assumed (stress relaxation for thin rods). Equation (6.41) can be rewritten as

$$\begin{aligned} T^{(1)} &= c^{(1)} S - e^{(1)} E, \\ D^{(1)} &= e^{(1)} S + \varepsilon^{(1)} E, \end{aligned} \tag{6.42}$$

where $c^{(1)}$, $e^{(1)}$, and $\varepsilon^{(1)}$ are the effective one-dimensional elastic, piezoelectric, and dielectric constants. They are related to the usual three-dimensional material constants through

$$\begin{aligned} c^{(1)} &= 1/s_{33}^{E}, \\ e^{(1)} &= d_{33}/s_{33}^{E}, \\ \varepsilon^{(1)} &= \varepsilon_{33}^{T} - d_{33}^{2}/s_{33}^{E}. \end{aligned} \tag{6.43}$$

Similarly, for the semiconductor,

$$\begin{aligned} T^{(2)} &= c^{(2)}S, \\ D^{(2)} &= \varepsilon^{(2)}E, \end{aligned} \tag{6.44}$$

where

$$\begin{aligned} c^{(2)} &= 1/s_{33}, \\ \varepsilon^{(2)} &= \varepsilon_{33}. \end{aligned} \tag{6.45}$$

For the currents in the semiconductor, we use the linearized constitutive relations with uniform doping:

$$\begin{aligned} J^{p} &\cong q p_{0} \mu^{p} E - q D^{p} \frac{\partial(\Delta p)}{\partial z}, \\ J^{n} &\cong q n_{0} \mu^{n} E + q D^{n} \frac{\partial(\Delta n)}{\partial z}, \end{aligned} \tag{6.46}$$

where

$$\begin{aligned} \mu^{p} &= \mu_{33}^{p}, \quad D^{p} = D_{33}^{p}, \\ \mu^{n} &= \mu_{33}^{n}, \quad D^{n} = D_{33}^{n}. \end{aligned} \tag{6.47}$$

For the composite rod, the total axial force can be calculated from

$$\begin{aligned} \widehat{T} &= T^{(1)}A^{(1)} + T^{(2)}A^{(2)} \\ &= \left(c^{(1)}S - e^{(1)}E\right)A^{(1)} + \left(c^{(2)}S\right)A^{(2)} \\ &= \left(c^{(1)}A^{(1)} + c^{(2)}A^{(2)}\right)S - e^{(1)}A^{(1)}E \\ &= \widehat{c}S - \widehat{e}E, \end{aligned} \tag{6.48}$$

where

$$\begin{aligned} \widehat{c} &= c^{(1)}A^{(1)} + c^{(2)}A^{(2)}, \\ \widehat{e} &= e^{(1)}A^{(1)}. \end{aligned} \tag{6.49}$$

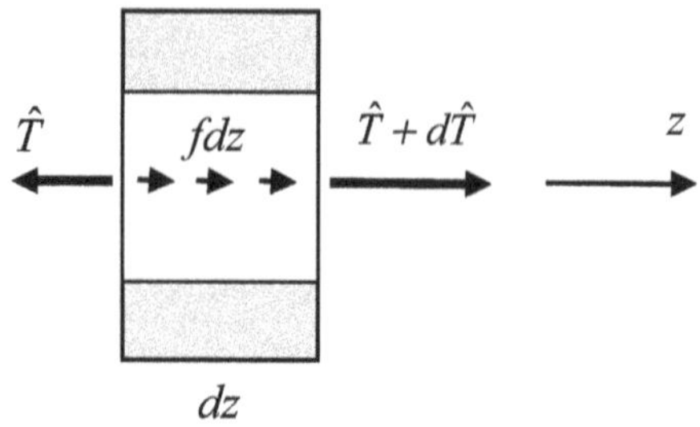

Fig. 6.3 A differential element of the composite rod

$A^{(1)}$ and $A^{(2)}$ are the cross sectional areas of the piezoelectric and semiconductor layers, respectively. Similarly, the total electric displacement over the cross section of the composite rod is

$$\begin{aligned}\widehat{D} &= D^{(1)}A^{(1)} + D^{(2)}A^{(2)} \\ &= \left(e^{(1)}S + \varepsilon^{(1)}E\right)A^{(1)} + \left(\varepsilon^{(2)}E\right)A^{(2)} \\ &= e^{(1)}A^{(1)}S + \left(\varepsilon^{(1)}A^{(1)} + \varepsilon^{(2)}A^{(2)}\right)E \\ &= \widehat{e}S + \widehat{\varepsilon}E,\end{aligned} \tag{6.50}$$

where

$$\widehat{\varepsilon} = \varepsilon^{(1)}A^{(1)} + \varepsilon^{(2)}A^{(2)}. \tag{6.51}$$

For extension, the equation of motion in the axial direction can be obtained by considering a differential element of the rod with length dz as shown in Fig. 6.3:

$$\frac{\partial \widehat{T}}{\partial z} + f(z,t) = 2b\left(\rho^{(1)}h + \rho^{(2)}c\right)\ddot{u}, \tag{6.52}$$

where $f(z,t)$ is the axial load per unit length of the rod. Similarly, the charge equation of electrostatics and the conservation of holes and electrons of the composite rod are

$$\frac{\partial \widehat{D}}{\partial z} = q(\Delta p - \Delta n)A^{(2)}, \tag{6.53}$$

$$\begin{aligned} q\frac{\partial}{\partial t}(\Delta p) &= -\frac{\partial J^p}{\partial z}, \\ q\frac{\partial}{\partial t}(\Delta n) &= \frac{\partial J^n}{\partial z}. \end{aligned} \tag{6.54}$$

We consider static extension of an electrically isolated rod. The boundary conditions are

$$\widehat{T}(\pm L) = F, \quad \widehat{D}(\pm L) = 0,$$
$$J^p(\pm L) = 0, \quad J^n(\pm L) = 0. \tag{6.55}$$

Δp and Δn must satisfy the following global charge neutrality conditions:

$$\int_{-L}^{L} \Delta p dz = 0, \quad \int_{-L}^{L} \Delta n dz = 0. \tag{6.56}$$

Only one of Eq. (6.56) is independent. The other is implied by integrating Eq. (6.53) between –L and L and using the electric displacement boundary conditions in Eq. (6.55), which leads to

$$\int_{-L}^{L} q(\Delta p - \Delta n) dz = 0. \tag{6.57}$$

To determine the mechanical displacement and the electric potential uniquely, we also set

$$u(0) = 0, \quad \varphi(0) = 0. \tag{6.58}$$

The polarization vector and effective polarization charge density are calculated from

$$P = D - \varepsilon_0 E = \frac{\widehat{D}}{A^{(1)} + A^{(2)}} - \varepsilon_0 E,$$
$$\rho^P = -P_{k,k} = -\frac{dP}{dz}. \tag{6.59}$$

The solution to the above boundary-value problem is given by

$$\varphi = \frac{\widehat{e}F}{\left(\widehat{\varepsilon}\widehat{c} + \widehat{e}^2\right) k \cosh kL} \sinh kz, \tag{6.60}$$

$$E = -\frac{\widehat{e}F}{\left(\widehat{\varepsilon}\widehat{c} + \widehat{e}^2\right) \cosh kL} \cosh kz, \tag{6.61}$$

$$D = \frac{\widehat{e}F}{\widehat{c}\left(A^{(1)} + A^{(2)}\right)} \left(1 - \frac{\cosh kz}{\cosh kL}\right), \tag{6.62}$$

$$p = p_0 - \frac{p_0 \mu^p}{D^p} \frac{\widehat{e}F}{\left(\widehat{\varepsilon}\widehat{c} + \widehat{e}^2\right) k \cosh kL} \sinh kz, \tag{6.63}$$

$$n = n_0 + \frac{n_0 \mu^n}{D^n} \frac{\hat{e}F}{\left(\widetilde{\varepsilon}\hat{c} + \hat{e}^2\right) k \cosh kL} \sinh kz, \tag{6.64}$$

$$P = \frac{\hat{e}F}{\hat{c}\cosh kL}\left(-\frac{1}{A^{(1)} + A^{(2)}} + \frac{\varepsilon_0}{\overline{\varepsilon}A^{(2)}}\right)\cosh kz + \frac{\hat{e}F}{\hat{c}\left(A^{(1)} + A^{(2)}\right)}, \tag{6.65}$$

$$\rho^P = \frac{\hat{e}Fk}{\hat{c}\cosh kL}\left(\frac{1}{A^{(1)} + A^{(2)}} + \frac{\varepsilon_0}{\overline{\varepsilon}A^{(2)}}\right)\sinh kz, \tag{6.66}$$

$$u = -\frac{\hat{e}^2 F}{\hat{c}\left(\widetilde{\varepsilon}\hat{c} + \hat{e}^2\right) k \cosh kL} \sinh kz + \frac{F}{\hat{c}} z, \tag{6.67}$$

$$S = -\frac{\hat{e}^2 F}{\hat{c}\left(\widetilde{\varepsilon}\hat{c} + \hat{e}^2\right) \cosh kL} \cosh kz + \frac{F}{\hat{c}}, \tag{6.68}$$

where

$$\begin{aligned} k^2 &= \left(\frac{p_0 \mu^p}{D^p} + \frac{n_0 \mu^n}{D^n}\right)\frac{q}{\widetilde{\varepsilon}}, \\ \widetilde{\varepsilon}A^{(2)} &= \hat{\varepsilon} + \frac{\hat{e}^2}{\hat{c}}. \end{aligned} \tag{6.69}$$

We are mainly interested in Δp and Δn produced by F. Consider Δn as an example. From Eq. (6.64) we obtain

$$\frac{\Delta n}{n_0} = \gamma \frac{F}{A^{(1)} + A^{(2)}} \frac{\sinh kz}{\cosh kL}, \tag{6.70}$$

where

$$\gamma = \frac{q}{k_B T} \frac{\hat{e}}{\widetilde{\varepsilon}\widetilde{c}k}, \quad \widetilde{c} = \frac{\hat{c} + \hat{e}^2/\hat{\varepsilon}}{A^{(1)} + A^{(2)}}. \tag{6.71}$$

$F/(A^{(1)} + A^{(2)})$ is the average extensional stress. $\widetilde{c}$ is an averaged, piezoelectrically stiffened elastic constant. At the end of a long rod where kL is large, $\sinh kL \cong \cosh kL$. Therefore γ describes the change of carrier concentration near the end of the rod per unit stress normalized by n_0 and can be used as a measure of the strength of the effect of F on Δn.

As a numerical example, consider a rod of PZT-5A and silicon doped into an n-type semiconductor with $n_0 = 10^{22}/\text{m}^3$. The geometric parameters are that $L = 6$ μm, $h = c = 0.05$ μm, and $b = 0.2$ μm. Figure 6.4 shows the effect of interest, i.e., mechanically induced redistribution of mobile charges in a composite rod of

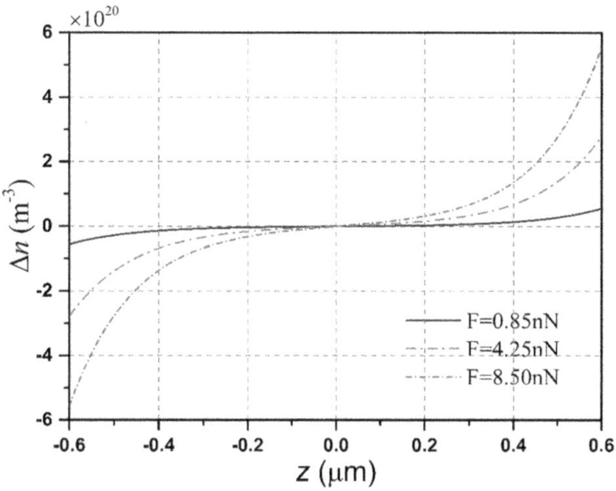

Fig. 6.4 Distribution of electron concentration perturbation Δn for different values of the axial force

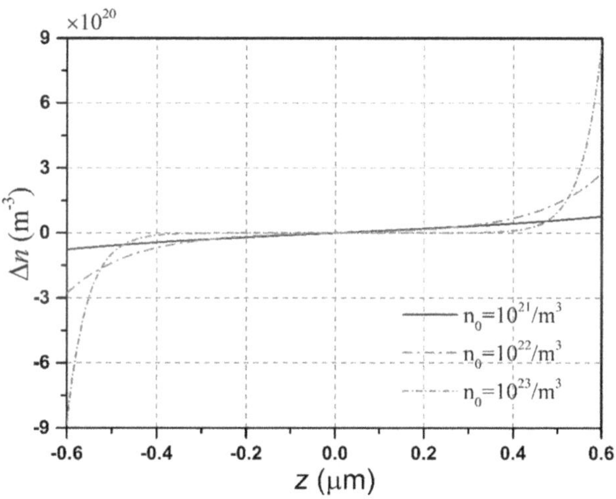

Fig. 6.5 Distribution of electron concentration perturbation Δn for different values of n_0

piezoelectric dielectrics and nonpiezoelectric semiconductors. Thus the composite rod effectively behaves like a homogeneous piezoelectric semiconductor rod such as ZnO.

Figure 6.5 shows the effect of the initial carrier concentration n_0 when $F = 4.25$ nN, while all other parameters are kept the same as those used for Fig. 6.4. It can be seen that the fields are sensitive to n_0. A larger n_0 causes a more rapid change of Δn near the ends of the rod. This can be explained by the dependence of k on n_0 in Eq. (6.69). A larger n_0 leads to a larger k and hence more rapid variations of the hyperbolic functions in Δn.

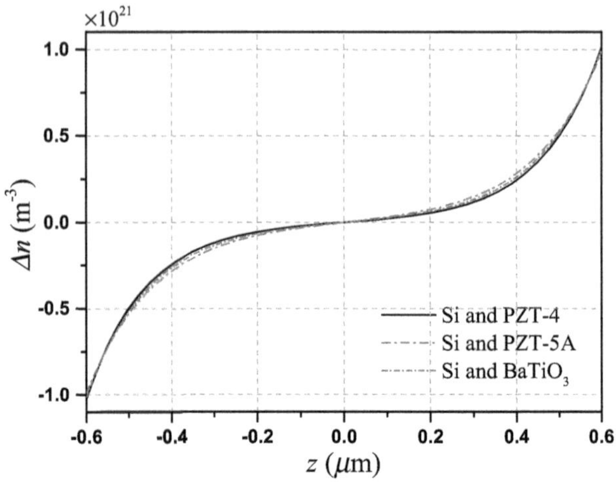

Fig. 6.6 Distribution of electron concentration perturbation Δn for different material combinations

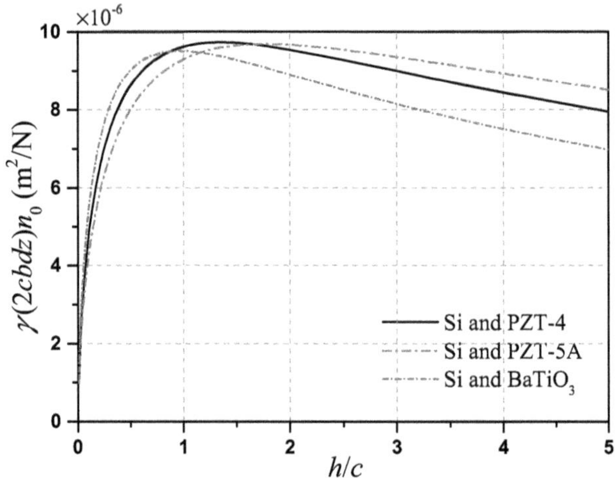

Fig. 6.7 $\gamma(2cbdz)n_0$ versus h/c for different material combinations

Figure 6.6 shows the effect of material combination when $n_0 = 10^{22}/\text{m}^3$ and $F = 4.25$ nN while all other parameters are kept the same as those used for Fig. 6.4. The results are very close.

The integration of Δn over $(0,L)$ can be used as a measure of the coupling between F and Δn as in Eq. (3.79). For the composite rod in this section, we are particularly interested in the effect of h/c rather than L on the coupling between F and Δn. Therefore we examine the change of number of carriers in the differential element in Fig. 6.3, i.e., $\Delta n(2cbdz)$. From Eq. (6.70), for a differential element of the rod near its ends under unit extensional stress, we have $\Delta n(2cbdz) \cong \gamma(2cbdz)n_0$ which is plotted in Fig. 6.7 versus h/c when $h + c$ is fixed. The curves all have a maximum near $h = c$, which is reasonable. When h is very small, the piezoelectric layers are very thin and cannot produce a lot of polarization charges to drive the mobile charges. When h is very large, the semiconductor layer is very thin, and there are not many mobile charges.

6.3 Bending of Beams with e_{33}

In this section we consider bending of the composite beam in Fig. 6.8 [4]. The beam consists of a nonpiezoelectric semiconductor layer such as Si and a pair of piezoelectric dielectric layers such as ceramics poled along the axial direction. The two ceramic layers are identical except that their poling directions are opposite to each other. The left end of the beam is fixed. The right end is under a transverse shear force F. The piezoelectric layers are unelectroded on their lateral surfaces. The electric field in the surrounding free space is neglected as usual as an approximation.

Consider bending without shear deformation in the x_3-x_2 plane. The flexural and axial displacements are approximately described by

$$u_2 \cong u_2(x_3, t), \quad u_3 \cong -x_2 u_{2,3}. \tag{6.72}$$

Thus the axial strain S_3 can be expressed in terms of u_2 as

$$S_3 = -x_2 u_{2,33}. \tag{6.73}$$

The electric potential and field are approximated by

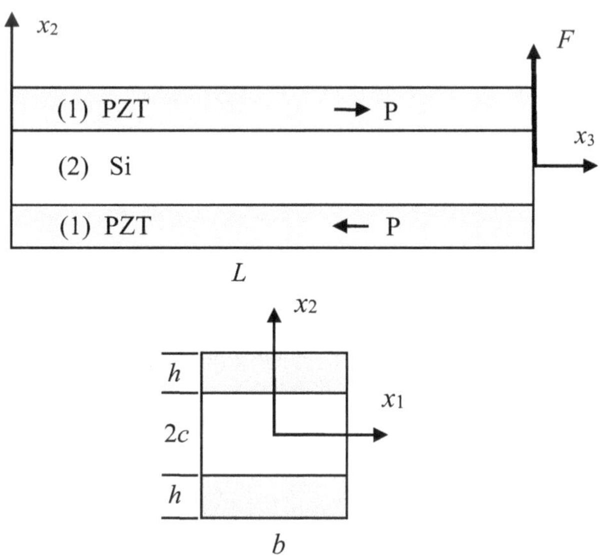

Fig. 6.8 Side view and cross section of a composite beam of piezoelectric dielectrics and nonpiezoelectric semiconductors

$$\varphi \cong \varphi(x_3,t), \ \ E_3 = -\varphi_{,3}. \tag{6.74}$$

Under the usual stress approximation for a beam in bending, we assume

$$T_1 \cong 0, \ \ T_2 \cong 0. \tag{6.75}$$

Then the relevant constitutive relations for the upper ceramic layer can be written as

$$\begin{aligned} S_3 &= s_{33}^{(1)} T_3^{(1)} + d_{33}^{(1)} E_3, \\ D_3^{(1)} &= d_{33}^{(1)} T_3^{(1)} + \varepsilon_{33}^{(1)} E_3. \end{aligned} \tag{6.76}$$

From Eq. (6.76) we solve for the stress and electric displacement to obtain

$$\begin{aligned} T_3^{(1)} &= \overline{c}_{33}^{(1)} S_3 - \overline{e}_{33}^{(1)} E_3, \\ D_3^{(1)} &= \overline{e}_{33}^{(1)} S_3 + \overline{\varepsilon}_{33}^{(1)} E_3, \end{aligned} \tag{6.77}$$

where

$$\begin{aligned} &\overline{c}_{33}^{(1)} = 1/s_{33}^{(1)}, \ \ \overline{e}_{33}^{(1)} = d_{33}^{(1)}/s_{33}^{(1)}, \\ &\overline{\varepsilon}_{33}^{(1)} = \varepsilon_{33}^{(1)}\left(1 - k_{33}^2\right), \ \ k_{33}^2 = \left(d_{33}^{(1)}\right)^2 / \left(\varepsilon_{33}^{(1)} s_{33}^{(1)}\right). \end{aligned} \tag{6.78}$$

Similarly, for the lower ceramic layer with opposite poling, formally Eqs. (6.76), (6.77), and (6.78) are still valid but $d_{33}^{(1)}$ changes its sign. For the semiconductor layer in the middle which is nonpiezoelectric, the constitutive relations are

$$\begin{aligned} T_3^{(2)} &= \overline{c}_{33}^{(2)} S_3 \\ D_3^{(2)} &= \varepsilon_{33}^{(2)} E_3, \end{aligned} \tag{6.79}$$

where $\overline{c}_{33}^{(2)} = 1/s_{33}^{(2)}$. The bending moment M is defined by the following integral over a cross section of the beam:

$$M = \int x_2 T_3 dx_1 dx_2 = -D u_{2,33} + 2G\overline{e}_{33}^{(1)} \varphi_{,3}, \tag{6.80}$$

where Eqs. (6.77) and (6.79) have been used, and

$$\begin{aligned} D &= \left\{ \frac{2}{3} \overline{c}_{33}^{(2)} c^3 + \frac{2}{3} \overline{c}_{33}^{(1)} \left[(c+h)^3 - c^3 \right] \right\} b, \\ G &= \left(c + \frac{h}{2} \right) hb. \end{aligned} \tag{6.81}$$

D is the bending stiffness of the beam. G is the first moment of the cross-sectional area of one of the ceramic layers about the x_1 axis. The transvers shear force Q in the beam is defined by the integration of T_{32} over a cross section:

$$Q = \int T_{32} dx_1 dx_2. \tag{6.82}$$

In elementary or classical bending without shear deformation, the constitutive relation for Q is not derived from its definition in Eq. (6.82). Instead, it will be provided by the moment-shear force relationship below. From the equation of motion of the differential element of the beam in Fig. 6.9 in the x_2 direction and its moment equation, we obtain

$$\begin{aligned} &Q_{,3} + f_2(x_3, t) = 2b\left(\rho^{(1)} h + \rho^{(2)} c\right) \ddot{u}_2, \\ &M_{,3} - Q = 0, \end{aligned} \tag{6.83}$$

where $f_2(x_3, t)$ is the transverse load per unit length of the beam. In Eq. $(6.83)_2$, the rotatory inertia is neglected and Eq. $(6.83)_2$ is the moment-shear force relationship needed. From Eq. $(6.83)_2$ and Eq. (6.80), we have

$$Q = -Du_{2,333} + 2G\overline{e}_{33}^{(1)} \varphi_{,33}. \tag{6.84}$$

The total electric displacement over a cross section of the beam is

$$\begin{aligned} \widehat{D}_3 &= \int D_3 dx_1 dx_2 \\ &= -2G\overline{e}_{33}^{(1)} u_{2,33} - \widehat{\varepsilon} \varphi_{,3}, \end{aligned} \tag{6.85}$$

where Eqs. (6.77) and (6.79) have been used, and

$$\begin{aligned} &\widehat{\varepsilon} = \overline{\varepsilon}_{33}^{(1)} A^{(1)} + \varepsilon_{33}^{(2)} A^{(2)}, \\ &A^{(1)} = 2bh, \quad A^{(2)} = 2bc. \end{aligned} \tag{6.86}$$

Similar to the derivation of Eq. (6.83), by considering the differential element in Fig. 6.9 under electrical loads, the charge equation of electrostatics can be written as

$$\widehat{D}_{3,3} = q(\Delta p - \Delta n) A^{(2)}. \tag{6.87}$$

For the one-dimensional currents, we use the linearized constitutive relations with uniform doping:

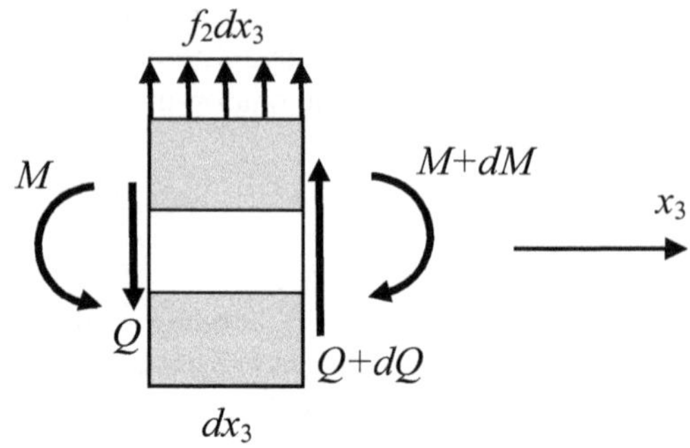

Fig. 6.9 A differential element of the beam under shear forces, bending moments, and transvers load

$$
\begin{aligned}
J_3^p &\cong qp_0\mu_{33}^p E_3 - qD_{33}^p(\Delta p)_{,3},\\
J_3^n &\cong qn_0\mu_{33}^n E_3 + qD_{33}^n(\Delta n)_{,3}.
\end{aligned}
\tag{6.88}
$$

The one-dimensional conservation of holes and electrons are

$$
\begin{aligned}
q\frac{\partial}{\partial t}(\Delta p) &= -J_{3,3}^p,\\
q\frac{\partial}{\partial t}(\Delta n) &= J_{3,3}^n.
\end{aligned}
\tag{6.89}
$$

We need to solve the boundary-value problem for the static bending of a cantilever beam under a constant end force F as shown in Fig. 6.8. The boundary conditions are

$$
\begin{aligned}
&u_2(0) = 0, \quad u_{2,3}(0) = 0, \quad \widehat{D}_3(0) = 0, \quad J_3^p(0) = 0, \quad J_3^n(0) = 0,\\
&M(L) = 0, \quad Q(L) = F, \quad \widehat{D}_3(L) = 0, \quad J_3^p(L) = 0, \quad J_3^n(L) = 0.
\end{aligned}
\tag{6.90}
$$

From Eqs. (6.83)$_1$, (6.84), and the mechanical boundary condition on the shear force at $x_3 = L$ in Eq. (6.90), we have

$$
-Du_{2,333} + 2G\overline{e}_{33}^{(1)}\varphi_{,33} = F. \tag{6.91}
$$

From Eq. (6.91) we solve for

$$
u_{2,333} = \frac{2G\overline{e}_{33}^{(1)}}{D}\varphi_{,33} - \frac{F}{D}. \tag{6.92}
$$

From Eq. (6.87), with the use of Eq. (6.85), we obtain

$$-2G\overline{e}_{33}^{(1)}u_{2,333}-\widehat{\varepsilon}\varphi_{,33}=q(\Delta p-\Delta n)A^{(2)}. \tag{6.93}$$

Equations (6.89), (6.88), and the current boundary conditions in Eq. (6.90) lead to

$$\begin{aligned}-qp_0\mu_{33}^p\varphi_{,3}-qD_{33}^p(\Delta p)_{,3}&=0,\\-qn_0\mu_3^n\varphi_{,3}+qD_{33}^n(\Delta n)_{,3}&=0.\end{aligned} \tag{6.94}$$

The integration of Eq. (6.94) gives

$$\begin{aligned}q\Delta p&=-\frac{qp_0\mu_{33}^p}{D_{33}^p}\varphi+C_1,\\-q\Delta n&=-\frac{qn_0\mu_{33}^n}{D_{33}^n}\varphi+C_2,\end{aligned} \tag{6.95}$$

where C_1 and C_2 are undetermined constants. The substitution of Eqs. (6.92) and (6.95) into Eq. (6.93) yields

$$\varphi_{,33}-k^2\varphi=\frac{2G\overline{e}_{33}^{(1)}}{\widetilde{\varepsilon}D}F-\frac{1}{\widetilde{\varepsilon}}(C_1+C_2)A^{(2)}, \tag{6.96}$$

where

$$\begin{aligned}k^2&=\frac{1}{\widetilde{\varepsilon}}\left(\frac{qp_0\mu_{33}^p}{D_{33}^p}+\frac{qn_0\mu_{33}^n}{D_{33}^n}\right)A^{(2)},\\\widetilde{\varepsilon}&=\widehat{\varepsilon}+\frac{\left(2G\overline{e}_{33}^{(1)}\right)^2}{D}.\end{aligned} \tag{6.97}$$

The general solution of Eq. (6.96) is

$$\begin{aligned}\varphi=&\,kC_3\sinh kx_3+kC_4\sinh k(x_3-L)\\&-\frac{2G\overline{e}_{33}^{(1)}}{k^2\widetilde{\varepsilon}D}F+\frac{1}{k^2\widetilde{\varepsilon}}(C_1+C_2)A^{(2)},\end{aligned} \tag{6.98}$$

where C_3 and C_4 are undetermined constants. Then, from Eq. (6.92),

$$\begin{aligned}u_2=&\,\frac{2G\overline{e}_{33}^{(1)}}{D}[C_3\cosh kx_3+C_4\cosh k(x_3-L)]\\&-\frac{F}{6D}x_3^3+C_5x_3^2+C_6x_3+C_7,\end{aligned} \tag{6.99}$$

where C_5 through C_7 are undetermined constants. We have five remaining boundary conditions in Eq. (6.90). To determine C_1 through C_7, we need two more conditions as follows. Δp and Δn must satisfy the following global charge neutrality conditions:

$$\int_0^L \Delta p dx_3 = 0, \quad \int_0^L \Delta n dx_3 = 0. \tag{6.100}$$

Only one of Eq. (6.100) is independent. The other is implied by integrating Eq. (6.87) between 0 and L and using $\widehat{D}(0) = \widehat{D}(L) = 0$, which gives

$$\int_0^L q(\Delta p - \Delta n) dx_3 = 0. \tag{6.101}$$

To determine the electric potential uniquely, we set

$$\varphi(0) = 0. \tag{6.102}$$

Equations (6.102) and one of Eq. (6.100) are the two additional conditions needed.

As an example, we consider PZTs and $BaTiO_3$ for the piezoelectric dielectric material and silicon for the nonpiezoelectric semiconductor. For geometric parameters we consider a beam with $L = 600$ nm, $h = 10$ nm, $c = 15$ nm, and $b = 50$ nm. The reference carrier concentrations are $p_0 = n_0 = 10^{23}/\text{m}^3$. The end force $F = 20 \times 10^{-12}$ N = 20 pN. Some of these parameters are varied separately in the figures below.

Figure 6.10 shows the main effect of interest, i.e., mechanically induced redistribution of mobile charges in the bending of a composite beam of piezoelectric dielectrics and nonpiezoelectric semiconductors. Thus the composite beam effectively behaves like a homogeneous piezoelectric semiconductor beam. Different from the static bending of a ZnO beam in Sect. 4.3 where the mobile charges are driven to the top and bottom surfaces of the beam, in the composite beam here the mobile charges are driven to the two ends of the beam. This is because of the specific orientations of the materials involved. The fields in the figure are linear in F. They become stronger or larger when F increases.

The change of number of carriers in the differential element in Fig. 6.9, i.e., Δp $(2cbdz)$ where $dz = dx_3$, may be used as a somewhat crude measure of the strength of the coupling between F and Δp. For a differential element of the beam near its right end, with $dz = 10$ nm, we plot $\Delta p(2cbdz)$ in Fig. 6.11 versus the thickness ratio h/c of the piezoelectric and semiconductor layers when $h + c$ is fixed. The curves for different material combinations all have a maximum roughly near $h = c$, which is reasonable.

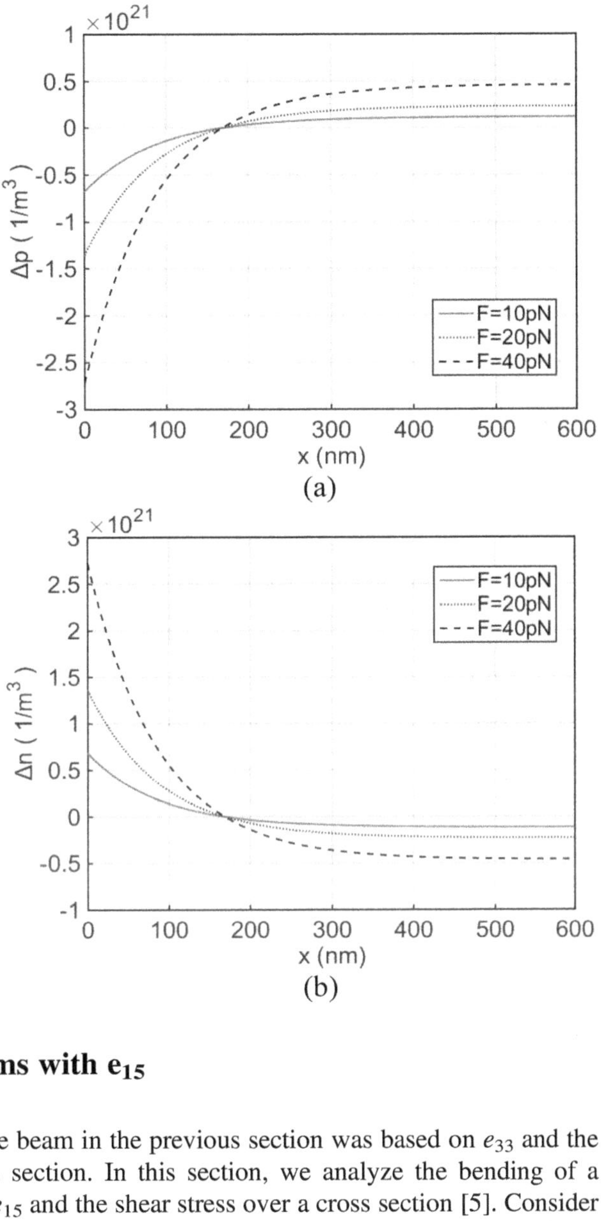

Fig. 6.10 Distributions of (**a**) hole concentration perturbation Δp, and (**b**) electron concentration perturbation Δn for different values of the shear force F in a composite beam of PZT-5A and silicon. $x = x_3$

6.4 Bending of Beams with e_{15}

The bending of a composite beam in the previous section was based on e_{33} and the normal stress over a cross section. In this section, we analyze the bending of a composite beam based on e_{15} and the shear stress over a cross section [5]. Consider the composite beam in Fig. 6.12. It consists of a piezoelectric dielectric middle layer labeled by "(1)" such as ceramics poled along the x_3 direction, and two identical nonpiezoelectric semiconductor layers labeled by "(2)" such as silicon at the top and bottom. Different from Fig. 6.8, the piezoelectric layer in Fig. 6.12 is in the middle because in bending the shear stress over a cross section is large in the middle. The left end of the beam is fixed. The right end is under a static transverse shear force F.

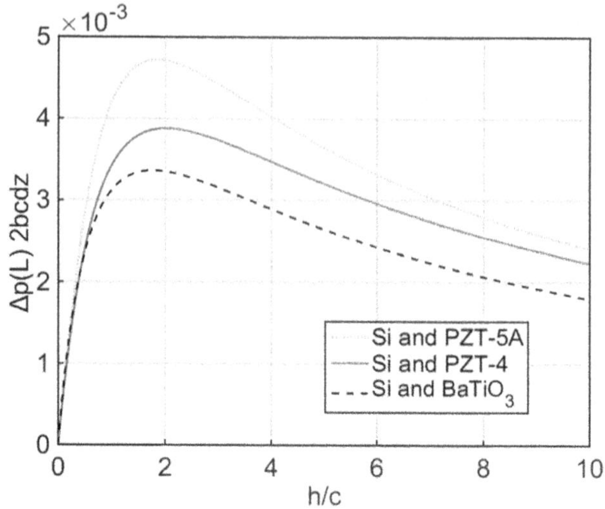

Fig. 6.11 $\Delta p(L) \times 2cbdz$ versus *h/c* for different material combinations when $p_0 = n_0 = 10^{23}/\text{m}^3$, $F = 20$ pN, $dz = 10$ nm, and $z = x_3$

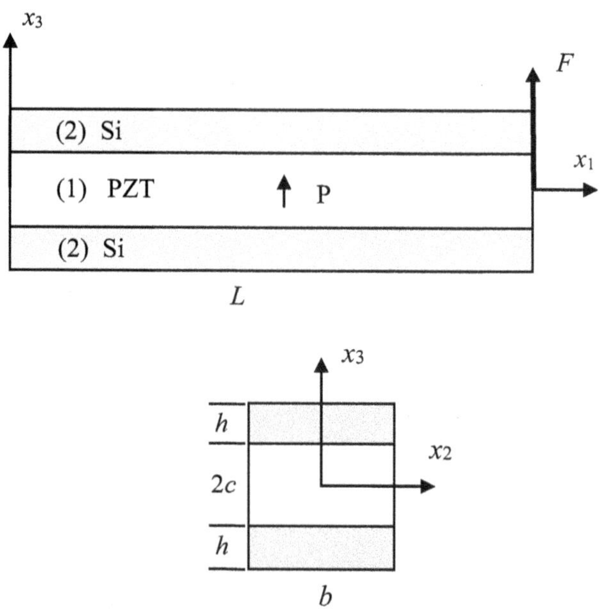

Fig. 6.12 Side view and cross section of a composite beam of piezoelectric dielectrics and nonpiezoelectric semiconductors

The piezoelectric layer is unelectroded on its lateral surfaces. The electric field in the surrounding free space is neglected as usual as an approximation.

To develop a one-dimensional model for bending in the x_1-x_3 plane with shear deformation, we make the following approximations of the relevant mechanical displacements and electric potential:

$$
\begin{aligned}
u_3(\mathbf{x},t) &\cong w(x_1,t), \\
u_1(\mathbf{x},t) &\cong x_3\psi(x_1,t), \\
\varphi(\mathbf{x},t) &\cong \varphi(x_1,t),
\end{aligned}
\tag{6.103}
$$

where $w(x_1,t)$ is the bending displacement (deflection) and $\psi(x_1,t)$ the shear deformation accompanying bending. The relevant strain and electric field components are

$$
\begin{aligned}
S_1 &= S_{11} = u_{1,1} = x_3\psi_{,1}, \\
S_5 &= 2S_{31} = u_{3,1} + u_{1,3} = w_{,1} + \psi, \\
E_1 &= -\varphi_{,1}.
\end{aligned}
\tag{6.104}
$$

For bending in the x_1-x_3 plane, the main stress components are the normal stress T_1 and shear stress $T_{13} = T_5$. We introduce the following stress relaxation for thin beams:

$$
T_2 = T_3 = T_4 = T_6 \cong 0. \tag{6.105}
$$

For the piezoelectric middle layer, the constitutive relations for the relevant strain and electric displacement components are

$$
\begin{aligned}
S_1 &= s_{11}^{(1)} T_1, \\
S_5 &= s_{55}^{(1)} T_5 + d_{15}E_1, \\
D_1 &= d_{15}T_5 + \varepsilon_{11}^{(1)} E_1.
\end{aligned}
\tag{6.106}
$$

We invert Eq. $(6.106)_{1,2}$ for expressions of stresses in terms of strains and substitute the resulting expression for T_5 into Eq. $(6.106)_3$. This gives

$$
\begin{aligned}
T_1 &= T_{11} = \overline{c}_{11}^{(1)} S_1 = \overline{c}_{11}^{(1)} x_3\psi_{,1}, \\
T_5 &= T_{31} = \overline{c}_{55}^{(1)} S_5 - \overline{e}_{15}E_1 = \overline{c}_{55}^{(1)}(w_{,1} + \psi) + \overline{e}_{15}\varphi_{,1}, \\
D_1 &= \overline{e}_{15}S_5 + \overline{\varepsilon}_{11}^{(1)} E_1 = \overline{e}_{15}(w_{,1} + \psi) - \overline{\varepsilon}_{11}^{(1)}\varphi_{,1},
\end{aligned}
\tag{6.107}
$$

where Eq. (6.104) has been used. The one-dimensional effective material constants for thin beams in Eq. (6.107) are

$$
\begin{aligned}
&\overline{c}_{11}^{(1)} = 1/s_{11}^{(1)}, \quad \overline{c}_{55}^{(1)} = 1/s_{55}^{(1)}, \quad \overline{e}_{15} = d_{15}/s_{55}^{(1)}, \\
&\overline{\varepsilon}_{11}^{(1)} = \varepsilon_{11}^{(1)} - d_{15}^2/s_{55}^{(1)}.
\end{aligned}
\tag{6.108}
$$

Similarly, for the semiconductor layers, the one-dimensional constitutive relations for thin beams are

$$\begin{aligned} T_1 &= \bar{c}_{11}^{(2)} S_1 = \bar{c}_{11}^{(2)} x_3 \psi_{,1}, \\ T_5 &= \bar{c}_{55}^{(2)} S_5 = \bar{c}_{55}^{(2)} (w_{,1} + \psi), \\ D_1 &= \bar{\varepsilon}_{11}^{(2)} E_1 = -\bar{\varepsilon}_{11}^{(2)} \varphi_{,1}, \end{aligned} \tag{6.109}$$

where

$$\bar{c}_{11}^{(2)} = 1/s_{11}^{(2)}, \quad \bar{c}_{55}^{(2)} = 1/s_{55}^{(2)}. \tag{6.110}$$

The carrier concentration perturbations in the semiconductor layers are approximated by

$$\begin{aligned} \Delta p(\mathbf{x}, t) &\cong \Delta p(x_1, t), \\ \Delta n(\mathbf{x}, t) &\cong \Delta n(x_1, t). \end{aligned} \tag{6.111}$$

The relevant one-dimensional linear constitutive relations for the current densities are

$$\begin{aligned} J_1^p &= q p_0 \mu_{11}^p E_1 - q D_{11}^p (\Delta p)_{,1} \\ &= -q p_0 \mu_{11}^p \varphi_{,1} - q D_{11}^p (\Delta p)_{,1}, \\ J_1^n &= q n_0 \mu_{11}^n E_1 + q D_{11}^n (\Delta n)_{,1} \\ &= -q n_0 \mu_{11}^n \varphi_{,1} + q D_{11}^n (\Delta n)_{,1}. \end{aligned} \tag{6.112}$$

The shear force Q, bending moment M, and total axial electric displacement $\widehat{D}$ are given by the following integrals over a typical cross section of the beam:

$$Q = \int T_{13} dx_2 dx_3 = \widehat{cA}(w_{,1} + \psi) + \bar{e}_{15} A^{(1)} \varphi_{,1}, \tag{6.113}$$

$$M = \int x_3 T_1 dx_1 dx_3 = \overline{D} \psi_{,1}, \tag{6.114}$$

$$\widehat{D} = \int D_1 dx_2 dx_3 = \bar{e}_{15} A^{(1)} (w_{,1} + \psi) - \widehat{\varepsilon A} \varphi_{,1}, \tag{6.115}$$

where

$$\begin{aligned} &A^{(1)} = 2bc, \quad A^{(2)} = 2bh, \quad \widehat{A} = A^{(1)} + A^{(2)}, \\ &\widehat{cA} = \bar{c}_{55}^{(1)} A^{(1)} + \bar{c}_{55}^{(2)} A^{(2)}, \quad \widehat{\varepsilon A} = \bar{\varepsilon}_{11}^{(1)} A^{(1)} + \varepsilon_{11}^{(2)} A^{(2)}, \\ &\overline{D} = \frac{2}{3} b c^3 \bar{c}_{11}^{(1)} + \frac{2}{3} b \left[(c+h)^3 - c^3 \right] \bar{c}_{11}^{(2)}. \end{aligned} \tag{6.116}$$

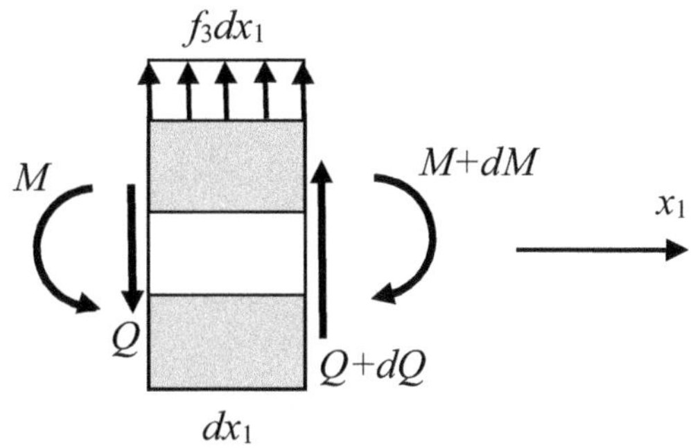

Fig. 6.13 A differential element of the beam under shear forces, bending moments, and transverse load

From the equation of motion of the differential element of the beam in Fig. 6.13 in the x_3 direction and its moment equation, we obtain

$$\begin{aligned} Q_{,1} + f_3(x_1, t) &= 2b\left(\rho^{(1)}h + \rho^{(2)}c\right)\ddot{u}_3, \\ M_{,1} - Q &= \bar{I}\ddot{\psi}, \end{aligned} \tag{6.117}$$

where $f_3(x_1,t)$ is the transverse load per unit length of the beam, and

$$\bar{I} = \frac{2}{3}bc^3\rho^{(1)} + \frac{2}{3}b\left[(c+h)^3 - c^3\right]\rho^{(2)}, \tag{6.118}$$

which is the rotatory inertia of the cross section about the x_2 axis. The effects of the rotatory inertia and shear deformation are more pronounced when the beams are not very long and thin. Similarly, by considering the differential element in Fig. 6.13 under electric loads, the charge equation of electrostatics can be written as

$$\hat{D}_{,1} = q(\Delta p - \Delta n)A^{(2)}. \tag{6.119}$$

The one-dimensional conservation of holes and electrons are

$$\begin{aligned} q\frac{\partial}{\partial t}(\Delta p) &= -J^p_{1,1}, \\ q\frac{\partial}{\partial t}(\Delta n) &= J^n_{1,1}. \end{aligned} \tag{6.120}$$

For a cantilever beam under a constant end force F as shown in Fig. 6.12, the boundary conditions are

$$\begin{aligned} &w(0) = 0, \quad \psi(0) = 0, \quad \widehat{D}(0) = 0, \quad J_1^p(0) = 0, \quad J_1^n(0) = 0, \\ &M(L) = 0, \quad Q(L) = F, \quad \widehat{D}(L) = 0, \quad J_1^p(L) = 0, \quad J_1^n(L) = 0. \end{aligned} \tag{6.121}$$

Δp and Δn must satisfy the following global charge neutrality conditions:

$$\int_0^L \Delta n dx_1 = 0, \quad \int_0^L \Delta p dx_1 = 0. \tag{6.122}$$

Only one of Eq. (6.122) is independent. To determine the electric potential uniquely, we set

$$\varphi(0) = 0. \tag{6.123}$$

Mathematically, we have a system of linear ordinary differential equations with constant coefficients. The solution can be obtained in a systematic manner. From Eq. (6.119), with the use of Eq. (6.115), we can write the charge equation as

$$\bar{e}_{15} A^{(1)} \left(w_{,11} + \psi_{,1}\right) - \widehat{\varepsilon A} \varphi_{,11} = q(\Delta p - \Delta n) A^{(2)}. \tag{6.124}$$

From Eqs. $(6.117)_1$ and (6.113), the bending equation for u_3 becomes

$$\widehat{cA}\left(w_{,11} + \psi_{,1}\right) + \bar{e}_{15} A^{(1)} \varphi_{,11} = 0. \tag{6.125}$$

Equations (6.112), (6.120) and the current boundary conditions in Eq. (6.121) lead to

$$\begin{aligned} -q p_0 \mu_{11}^p \varphi_{,1} - q D_{11}^p (\Delta p)_{,1} = 0, \\ -q n_0 \mu_{11}^p \varphi_{,1} + q D_{11}^p (\Delta n)_{,1} = 0. \end{aligned} \tag{6.126}$$

Integrating Eqs. (6.126), we obtain

$$\begin{aligned} q\Delta p &= -\frac{q p_0 \mu_{11}^p \varphi}{D_{11}^p} + C_1, \\ -q\Delta n &= -\frac{q n_0 \mu_{11}^n \varphi}{D_{11}^n} + C_2, \end{aligned} \tag{6.127}$$

where C_1 and C_2 are undetermined constant. With the use of Eqs. (6.125) and (6.127), we can write the charge equation in Eq. (6.124) as

$$\varphi_{,11} - k^2 \varphi = -\frac{(C_1 + C_2) A^{(2)}}{\widetilde{\varepsilon}}, \tag{6.128}$$

where

$$k^2 = \left(\frac{qp_0\mu_{11}^p}{D_{11}^p} + \frac{qn_0\mu_{11}^n}{D_{11}^n}\right)\frac{A^{(2)}}{\widetilde{\varepsilon}},$$
$$\widetilde{\varepsilon} = \frac{\left(\overline{e}_{15}A^{(1)}\right)^2}{\widehat{cA}} + \widehat{\varepsilon A}. \tag{6.129}$$

The general solution of Eq. (6.128) is

$$\varphi = kC_3 \sinh kx_1 + kC_4 \cosh kx_1 + \frac{(C_1 + C_2)A^{(2)}}{k^2\widetilde{\varepsilon}}, \tag{6.130}$$

where C_3 and C_4 are undetermined constants. From Eqs. (6.114), (6.117)$_2$, and the mechanical boundary condition on the shear force at $x_3 = L$ in Eq. (6.121), the moment equation takes the following form:

$$\overline{D}\psi_{,11} = F, \tag{6.131}$$

which implies that

$$\psi = \frac{1}{2}\frac{F}{\overline{D}}x_1^2 + C_5x_1 + C_6, \tag{6.132}$$

where C_5 and C_6 are undetermined constants. From $Q = F$ and Eq. (6.113), we obtain

$$w = \frac{Fx_1 - \overline{e}_{15}A^{(1)}\varphi + C_7}{\widehat{cA}} - \int \psi dx_1, \tag{6.133}$$

where C_7 is an integration constant. To determine C_1 through C_7, we need to use the remaining ones of Eq. (6.121), one of Eq. (6.122) and Eq. (6.123).

For a numerical example consider a beam of PZT4 and silicon. We choose the following geometric parameters that $L = 600$ nm, $h = 10$ nm, $c = 15$ nm, and $b = 50$ nm. $p_0 = n_0 = 10^{23}$ m^{-3}. The basic fields of interest are shown in Fig. 6.14 for different values of the end shear force F. Although the shear force $Q = F$ is a constant, the shear strain in (a) varies a little along the beam, especially near the two ends. This is related to the fact that the electric potential and field in the beam change near the two ends and sometimes significantly as shown in (b) because of the hyperbolic functions in Eq. (6.130). Since the axial electric field E_1 is negative, the holes are driven to the left and the electrons to the right in (c) and (d). This is similar to the composite beam in the previous section but different from the homogeneous beam in Sect. 4.3 where the charges are driven to the top and bottom of the beam. For all fields, a larger F produces stronger fields as expected. Similar to the previous section, there also exists an optimal thickness ratio between the two types of layers for strong coupling between bending and mobile charge redistribution.

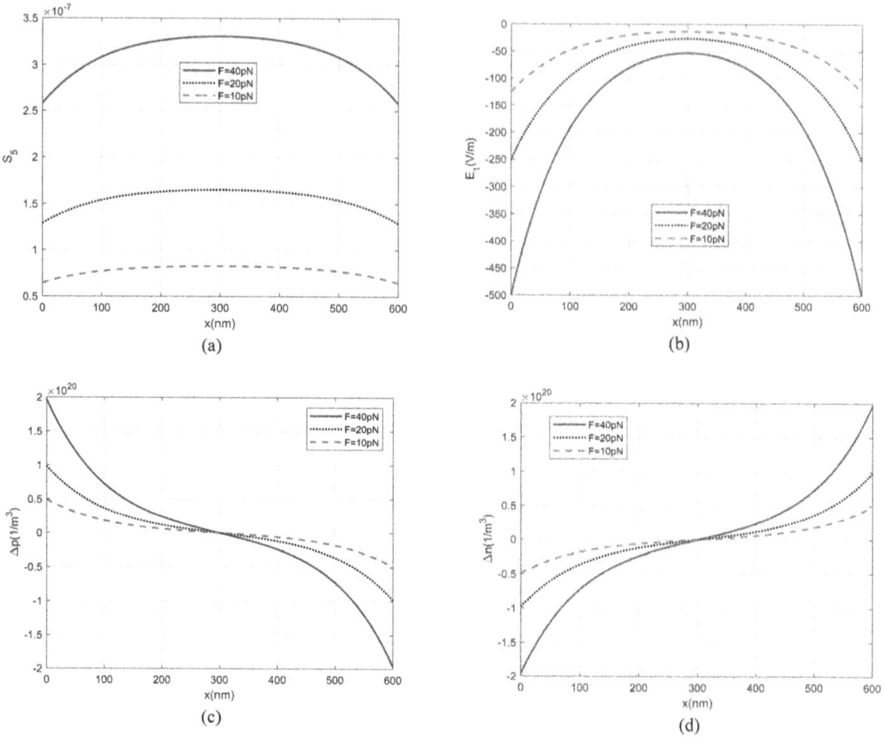

Fig. 6.14 Fields for different values of F. (**a**) Shear strain S_5. (**b**) Axial electric field E_1. (**c**) Hole concentration perturbation. (**d**) Electron concentration perturbation

6.5 Bending of a PN Junction

This section is concerned with PN junctions and bending of beams [6]. We show that PN junctions with coupling to bending deformation can be realized in properly constructed composite piezoelectric semiconductor beams. Figure 6.15a shows a composite beam with a PN junction. For a homogeneous junction the material is uniform along the beam but the two halves are doped differently. The left half has more holes and the right half more electrons. For a heterogeneous junction the materials in the two halves are different. (b) shows the doping profile of an abrupt junction.

We use the one-dimensional equations derived in Sect. 6.3. The field equations are from Eqs. (6.83)$_1$, (6.87), and (6.89):

$$\begin{aligned} Q_{,3} &= 2b\left(\rho^{(1)}h + \rho^{(2)}c\right)\ddot{u}_2, \\ \widehat{D}_{3,3} &= q(\Delta p - \Delta n)A^{(2)}, \end{aligned} \qquad (6.134)$$

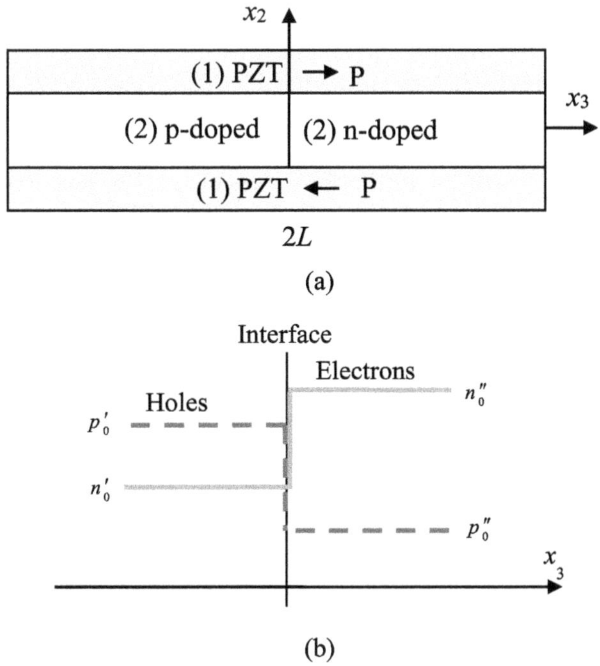

Fig. 6.15 (**a**) A PN junction in a composite beam. (**b**) Doping profile

$$
\begin{aligned}
q\frac{\partial}{\partial t}(\Delta p) &= -J^p_{3,3},\\
q\frac{\partial}{\partial t}(\Delta n) &= J^n_{3,3}.
\end{aligned}
\tag{6.135}
$$

The one-dimensional constitutive relations are from Eqs. (6.80), (6.84), (6.85), and (6.88):

$$
\begin{aligned}
M &= -Du_{2,33} - 2G\overline{e}^{(1)}_{33}E_3,\\
Q &= -Du_{2,333} - 2G\overline{e}^{(1)}_{33}E_{,3},\\
\widehat{D}_3 &= -2G\overline{e}^{(1)}_{33}u_{2,33} + \widehat{\varepsilon}E_3,
\end{aligned}
\tag{6.136}
$$

$$
\begin{aligned}
J^p_3 &\cong qp_0\mu^p_{33}E_3 - qD^p_{33}(\Delta p)_{,3},\\
J^n_3 &\cong qn_0\mu^n_{33}E_3 + qD^n_{33}(\Delta n)_{,3},
\end{aligned}
\tag{6.137}
$$

where, from Eq. (6.74),

$$
E_3 = -\varphi_{,3}. \tag{6.138}
$$

The axial electric polarization P_3 and the related effective polarization charge density ρ^P can be calculated from the following expressions:

$$
\begin{aligned}
P_3 &= D_3 - \varepsilon_0 E_3, \\
D_3 &= \widehat{D}_3 / \left(A^{(1)} + A^{(2)}\right), \\
\rho^P &= -P_{3,3}.
\end{aligned} \tag{6.139}
$$

We are interest in the formation of the PN junction through diffusion and the associated electromechanical fields. It is sufficient to consider the static bending of a free beam whose bending moment M and the shear force Q are identically zero along the beam. In this case, Eqs. (6.92), (6.93), (6.95), and (6.96) take the following form:

$$
u_{2,33} = \frac{2G\overline{e}_{33}^{(1)}}{D} \varphi_{,3}, \tag{6.140}
$$

$$
-2G\overline{e}_{33}^{(1)} u_{2,333} - \widehat{\varepsilon}\varphi_{,33} = q(\Delta p - \Delta n)A^{(2)}, \tag{6.141}
$$

$$
\begin{aligned}
q\Delta p &= -\frac{q p_0 \mu_{33}^p}{D_{33}^p} \varphi + C_1, \\
-q\Delta n &= -\frac{q n_0 \mu_{33}^n}{D_{33}^n} \varphi + C_2,
\end{aligned} \tag{6.142}
$$

$$
\varphi_{,33} - k^2 \varphi = -\frac{1}{\widetilde{\varepsilon}}(C_1 + C_2)A^{(2)}, \tag{6.143}
$$

where

$$
\begin{aligned}
k^2 &= \frac{1}{\widetilde{\varepsilon}} \left(\frac{q p_0 \mu_{33}^p}{D_{33}^p} + \frac{q n_0 \mu_{33}^n}{D_{33}^n} \right) A^{(2)}, \\
\widetilde{\varepsilon} &= \widehat{\varepsilon} + \frac{\left(2G\overline{e}_{33}^{(1)}\right)^2}{D},
\end{aligned} \tag{6.144}
$$

and C_1 and C_2 are undetermined constants. We need to treat the two halves of the beam in Fig. 6.15a separately. We use a prime for the material parameters and undetermined constants of the left half and a double prime for those of the right half. For $x_3 < 0$, we write

$$
\varphi = k' C'_3 \exp(k' x_3) - k' C'_4 \exp(-k' x_3) + \frac{1}{(k')^2 \widetilde{\varepsilon}'} (C'_1 + C'_2)A^{(2)}, \tag{6.145}
$$

$$
u_2 = \frac{2G\overline{e}_{33}^{(1)}}{D} \left[C'_3 \exp(k' x_3) + C'_4 \exp(-k' x_3) \right] + C'_5 x_3 + C'_6, \tag{6.146}
$$

$$
\begin{aligned}
q\Delta p &= -\frac{q p'_0 \mu^p_{33}}{D^p_{33}}\varphi + C'_1, \\
-q\Delta n &= -\frac{q n'_0 \mu^n_{33}}{D^n_{33}}\varphi + C'_2.
\end{aligned} \tag{6.147}
$$

Similarly, for $x_3 > 0$, we have

$$
\varphi = k''C''_3 \exp(k''x_3) - k''C''_4 \exp(-k''x_3) + \frac{1}{(k'')^2 \widetilde{\varepsilon}''}(C''_1 + C''_2)A^{(2)}, \tag{6.148}
$$

$$
u_2 = \frac{2G\overline{e}^{(1)}_{33}}{D}\left[C''_3 \exp(k''x_3) + C''_4 \exp(-k''x_3)\right] + C''_5 x_3 + C''_6, \tag{6.149}
$$

$$
\begin{aligned}
q\Delta p &= -\frac{q p''_0 \mu^p_{33}}{D^p_{33}}\varphi + C''_1, \\
-q\Delta n &= -\frac{q n''_0 \mu^n_{33}}{D^n_{33}}\varphi + C''_2.
\end{aligned} \tag{6.150}
$$

The boundary conditions at the two ends of the beam and the continuity conditions at the junction interface are

$$
\begin{aligned}
&M(\pm L) = 0, \quad Q(\pm L) = 0, \\
&\widehat{D}_3(\pm L) = 0, \\
&J^p_3(\pm L) = 0, \quad J^n_3(\pm L) = 0,
\end{aligned} \tag{6.151}
$$

$$
\begin{aligned}
&u_0(0^-) = u_2(0^+), \quad u_{2,3}(0^-) = u_{2,3}(0^+), \\
&M(0^-) = M(0^+), \quad Q(0^-) = Q(0^+), \\
&\varphi(0^-) = \varphi(0^+), \quad \widehat{D}_3(0^-) = \widehat{D}_3(0^+) \\
&p(0^-) = p(0^+), \quad n(0^-) = n(0^+), \\
&J^p_3(0^-) = J^p_3(0^+), \quad J^n_3(0^-) = J^n_3(0^+).
\end{aligned} \tag{6.152}
$$

To make the displacements and electric potential unique, we impose

$$
u_2(0) = 0, \; u_{2,3}(0) = 0, \; \varphi(0) = 0. \tag{6.153}
$$

We also have the following charge neutrality conditions:

$$
\int_{-L}^{L} \Delta p dx_3 = 0, \quad \int_{-L}^{L} \Delta n dx_3 = 0. \tag{6.154}
$$

Only one of Eq. (6.154) is independent. The substitution of Eqs. (6.145), (6.146), (6.147), (6.148), (6.149), and (6.150) into Eqs. (6.151), (6.152), (6.153), and (6.154) yields a system of linear equations for the undetermined constants. These equations are solved on a computer using MATLAB. For a complete description of the bending deformation, we introduce the bending curvature

$$K = \frac{u_{2,33}}{\left[1 + (u_{2,3})^2\right]^{3/2}} \cong u_{2,33}, \tag{6.155}$$

which has been approximated according the usual small deformation assumption in mechanics of materials.

As a numerical example, consider a beam with $L = 20 \times 10^{-6}$ m, $b = 50 \times 10^{-9}$ m, $c = 15 \times 10^{-9}$ m, and $h = 10 \times 10^{-9}$ m. Let

$$\begin{aligned} p_0' &= n_0'' = a_0 + \lambda a_0, \\ n_0' &= p_0'' = 0.7a_0 - \lambda a_0, \\ a_0 &= 10^{23}\ \text{m}^{-3}. \end{aligned} \tag{6.156}$$

When λ assumes different values, we have different doping levels.

Figure 6.16 shows the distributions of p and n along the beam near the junction. They have become smooth distributions because of diffusion (compare to Fig. 6.15b).

Figure 6.17 shows the bending deformation associated with the free PN junction. The coupling to bending offers the possibility of constructing new piezotronic devices operating with bending of composite beams.

For heterogeneous junctions we consider a relatively simple case by reversing the poling directions in the ceramics layers of the left half of the beam in Fig. 6.15a. When the poling direction is reversed, the relevant piezoelectric constants change their signs. The deflection curve and bending curvature associated with such a heterogeneous junction are shown in Fig. 6.18 for different doping levels. Compared to the corresponding fields of a homogeneous junction, the deflection curve and the bending curvature change their signs across the heterogeneous junction. The electric fields are qualitatively similar to those of the homogeneous junction and therefore are not shown.

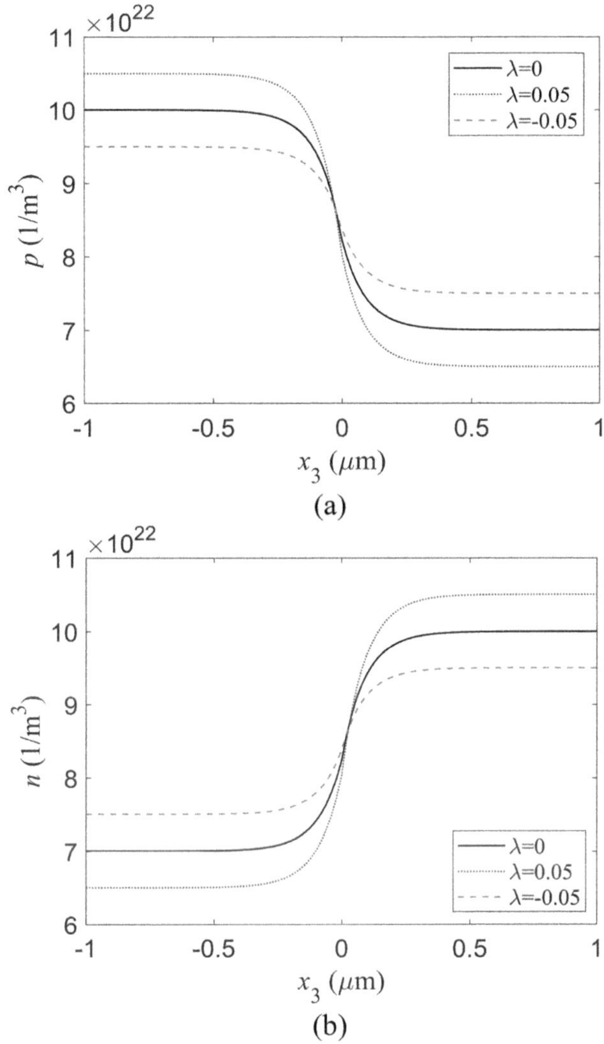

Fig. 6.16 Carrier concentration distributions near the homogeneous junction. (**a**) Holes. (**b**) Electrons

Fig. 6.17 Mechanical fields at the homogeneous junction. (**a**) Deflection curve. (**b**) Bending curvature

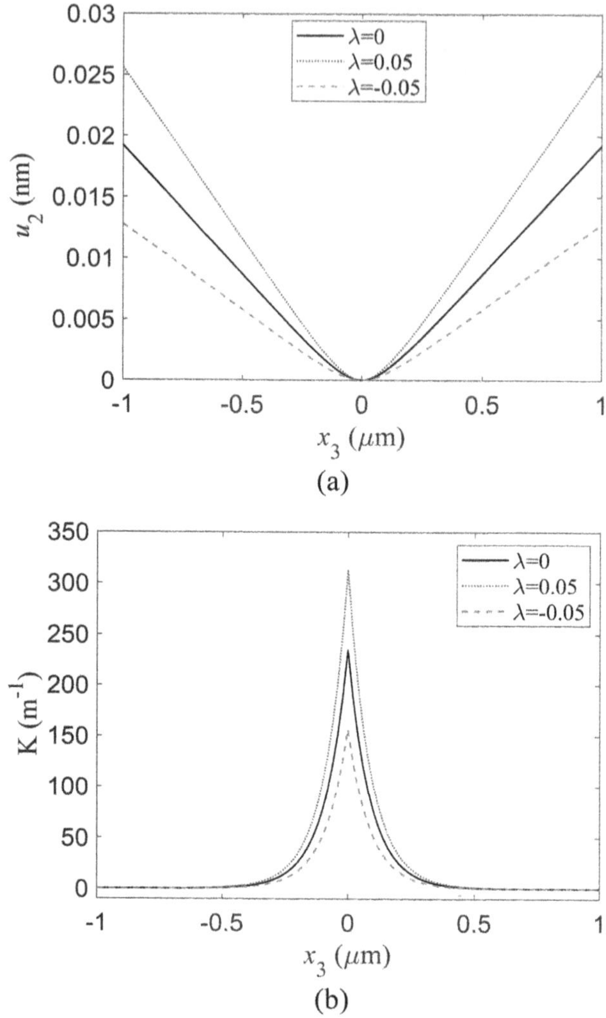

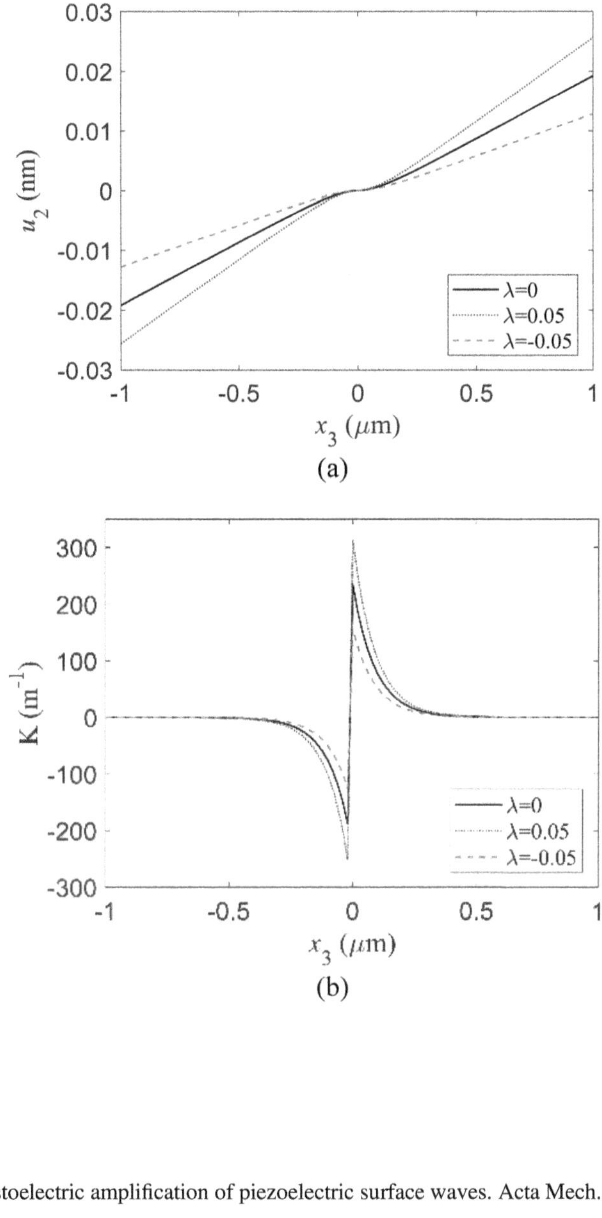

Fig. 6.18 Mechanical fields at the heterogeneous junction. (**a**) Deflection curve. (**b**) Bending curvature

References

1. J.S. Yang, H.G. Zhou, Acoustoelectric amplification of piezoelectric surface waves. Acta Mech. **172**, 113–122 (2004)
2. J.L. Bleustein, A new surface wave in piezoelectric materials. Appl. Phys. Lett. **13**, 412–413 (1968)
3. R.R. Cheng, C.L. Zhang, W.Q. Chen, J.S. Yang, Piezotronic effects in the extension of a composite fiber of piezoelectric dielectrics and nonpiezoelectric semiconductors. J. Appl. Phys. **124**, 064506 (2018)

4. Y.X. Luo, C.L. Zhang, W.Q. Chen, J.S. Yang, Piezopotential in a bended composite fiber made of a semiconductive core and of two piezoelectric layers with opposite polarities. Nano Energy **54**, 341–348 (2018)
5. K. Fang, Z.H. Qian, J.S. Yang, Piezopotential in a composite cantilever of piezoelectric dielectrics and nonpiezoelectric semiconductors produced by shear force through e_{15}. Mater. Res. Express **6**, 115917 (2019)
6. G.Y. Yang, L. Yang, J.K. Du, J. Wang, J.S. Yang, PN junctions with coupling to bending deformation in composite piezoelectric semiconductor fibers. Int. J. Mech. Sci. **1**, 105421 (2020)

Chapter 7
Thermal Effects

In this chapter we study thermal effects in piezoelectric semiconductor structures through pyroelectric and thermoelastic couplings. Both homogeneous and composite structures are considered.

7.1 Extension of Homogeneous Rods

Consider the free piezoelectric semiconductor rod shown in Fig. 7.1. The rod is within $|x_3| < L$. The c-axis of the (6mm) crystal is along the x_3 axis. There are no electromechanical loads. We study the electromechanical fields in the rod associated with the extensional deformation under a uniform temperature change [1].

Consider a uniform and known temperature change denoted by $\theta = \Theta - \Theta_0$ from a uniform reference temperature Θ_0. At Θ_0, the carrier concentrations are $p_0 = N_A^-$ and $n_0 = N_D^+$ which are assumed to be constants. Under the given temperature change, the perturbations of the carrier concentrations are denoted by Δp and Δn. For convenience we collect the relevant equations from Sect. 1.5. They are valid for the rest of the chapter. Since the temperature change is known, the heat equation in Eq. $(1.43)_3$ is not needed. The electromechanical fields are governed by

$$T_{ji,j} = \rho \ddot{u}_i, \tag{7.1}$$

$$D_{i,i} = q(\Delta p - \Delta n), \tag{7.2}$$

$$\begin{aligned} J^p_{i,i} &= -q\frac{\partial(\Delta p)}{\partial t}, \\ J^n_{i,i} &= q\frac{\partial(\Delta n)}{\partial t}, \end{aligned} \tag{7.3}$$

J. Yang, *Analysis of Piezoelectric Semiconductor Structures*,
https://doi.org/10.1007/978-3-030-48206-0_7

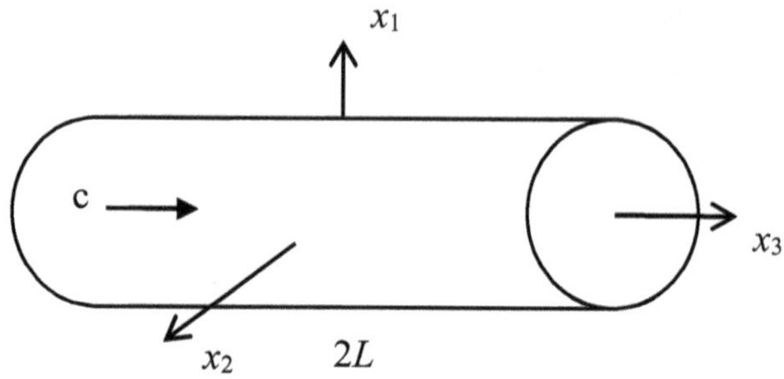

Fig. 7.1 A free piezoelectric semiconductor rod

$$S_{ij} = s^E_{ijkl} T_{kl} + d_{kij} E_k + \alpha_{ij} \theta,$$
$$D_i = d_{ikl} T_{kl} + \varepsilon^T_{ik} E_k + p_i \theta, \tag{7.4}$$

$$J^p_i \cong q p_0 \mu^p_{ij} E_j - q D^p_{ij} (\Delta p)_{,j},$$
$$J^n_i \cong q n_0 \mu^n_{ij} E_j + q D^n_{ij} (\Delta n)_{,j}. \tag{7.5}$$

In Eq. (7.5), thermoelectric couplings disappear because of the uniform temperature.

For a thin rod, we have, approximately

$$u_3 \cong u_3(x_3, t), \quad \varphi \cong \varphi(x_3, t), \tag{7.6}$$

which are uniform over a cross section. We also assume

$$\Delta p \cong \Delta p(x_3, t), \; \Delta n \cong \Delta n(x_3, t). \tag{7.7}$$

From Eq. (7.6), the axial strain and electric field are

$$S_3 = S_{33} = u_{3,3}, \; E_3 = -\varphi_{,3}. \tag{7.8}$$

For thin rods the stress is approximately uniaxial, i.e.,

$$T_1 = T_2 \cong 0. \tag{7.9}$$

Substituting Eq. (7.9) into the relevant ones of Eq. (7.4) and solving for T_{33} and D_3, we obtain

$$T_{33} = \bar{c}_{33} S_{33} - \bar{e}_{33} E_3 - \bar{\lambda}_{33} \theta,$$
$$D_3 = \bar{e}_{33} S_{33} + \bar{\varepsilon}_{33} E_3 + \bar{p}_3 \theta, \tag{7.10}$$

where the effective one-dimensional material constants are defined by

$$
\begin{aligned}
&\overline{c}_{33} = 1/s_{33}^{E}, \quad \overline{e}_{33} = d_{33}/s_{33}^{E}, \\
&\overline{\varepsilon}_{33} = \varepsilon_{33}^{T} - d_{33}^{2}/s_{33}^{E}, \quad \overline{\lambda}_{33} = \alpha_{33}/s_{33}^{E}, \\
&\overline{p}_{3} = p_{3} - d_{33}\alpha_{33}/s_{33}^{E}.
\end{aligned} \tag{7.11}
$$

The one-dimensional constitutive relations for the currents are

$$
\begin{aligned}
J_{3}^{p} &\cong qp_{0}\mu_{33}^{p}E_{3} - qD_{33}^{p}(\Delta p)_{,3}, \\
J_{3}^{n} &\cong qn_{0}\mu_{3}^{n}E_{3} + qD_{33}^{n}(\Delta n)_{,3}.
\end{aligned} \tag{7.12}
$$

We consider static extension. From Eqs. (7.1), (7.2), and (7.3), the one-dimensional field equations are

$$
\begin{aligned}
T_{33,3} &= 0, \\
D_{3,3} &= q(\Delta p - \Delta n),
\end{aligned} \tag{7.13}
$$

$$
\begin{aligned}
J_{3,3}^{p} &= 0, \\
J_{3,3}^{n} &= 0.
\end{aligned} \tag{7.14}
$$

The substitution of Eqs. (7.10) and (7.12) into Eqs. (7.13) and (7.14) gives four second-order equations for u_3, φ, Δp, and Δn. For an electromechanically isolated rod, the boundary conditions are

$$
\begin{aligned}
&T_{33}(\pm L) = 0, \quad D_{3}(\pm L) = 0, \\
&J^{n}(\pm L) = 0, \quad J^{p}(\pm L) = 0.
\end{aligned} \tag{7.15}
$$

The four equations for u_3, φ, Δp, and Δn are linear ordinary differential equations with constant coefficient. Their general solution can be obtained in a straightforward manner as

$$
\begin{aligned}
u_{3} &= \frac{\overline{e}_{33}}{\overline{c}_{33}}\frac{q}{k^{2}\widehat{\varepsilon}_{33}}\{A_{1}\sinh[k(x_{3}-L)] + A_{2}\sinh[k(x_{3}+L)]\} \\
&\quad + A_{5}x_{3} + A_{6}, \\
\varphi &= -\frac{q}{k^{2}\widehat{\varepsilon}_{33}}\{A_{1}\sinh[k(x_{3}-L)] + A_{2}\sinh[k(x_{3}+L)]\} \\
&\quad + A_{3}x_{3} + A_{4},
\end{aligned} \tag{7.16}
$$

$$
\begin{aligned}
\Delta p &= p_{0}\frac{q}{k_{B}\Theta_{0}}\frac{q}{k^{2}\widehat{\varepsilon}_{33}}\{A_{1}\sinh[k(x_{3}-L)] + A_{2}\sinh[k(x_{3}+L)]\} \\
&\quad + A_{7}x_{3} + A_{8}, \\
\Delta n &= -n_{0}\frac{q}{k_{B}\Theta_{0}}\frac{q}{k^{2}\widehat{\varepsilon}_{33}}\{A_{1}\sinh[k(x_{3}-L)] + A_{2}\sinh[k(x_{3}+L)]\} \\
&\quad + A_{7}x_{3} + A_{8},
\end{aligned} \tag{7.17}
$$

where

$$
\begin{aligned}
k^2 &= (p_0 + n_0)\frac{q}{k_B\Theta_0}\frac{q}{\widehat{\varepsilon}_{33}},\\
\widehat{\varepsilon}_{33} &= \overline{\varepsilon}_{33} - \frac{\overline{e}_{33}\overline{e}_{33}}{\overline{c}_{33}} = \varepsilon_{33}.
\end{aligned}
\tag{7.18}
$$

In Eqs. (7.16) and (7.17), A_1 through A_8 are undetermined constants. It can be shown that only five of the eight boundary conditions in Eq. (7.15) are independent. We need to impose the global charge neutrality conditions:

$$
\int_{-L}^{L} \Delta p dx_3 = 0, \quad \int_{-L}^{L} \Delta n dx_3 = 0,
\tag{7.19}
$$

among which only one is independent. To determine the displacement and potential fields uniquely, we set

$$
u_3(0) = 0, \quad \varphi(0) = 0.
\tag{7.20}
$$

Thus there are eight independent conditions for determining the eight unknown constants. The final solution is

$$
\begin{aligned}
u_3 &= -\frac{\overline{e}_{33}G_0\theta}{\overline{c}_{33}k}\frac{\sinh kx_3}{\cosh kL} + \frac{\overline{\lambda}_{33}\theta}{\overline{c}_{33}}x_3,\\
\varphi &= \frac{G_0\theta}{k}\frac{\sinh kx_3}{\cosh kL},
\end{aligned}
\tag{7.21}
$$

$$
\begin{aligned}
\Delta p &= -p_0\frac{q}{k_B\Theta_0}\frac{G_0\theta}{k}\frac{\sinh kx_3}{\cosh kL},\\
\Delta n &= n_0\frac{q}{k_B\Theta_0}\frac{G_0\theta}{k}\frac{\sinh kx_3}{\cosh kL},
\end{aligned}
\tag{7.22}
$$

where

$$
G_0 = \frac{\overline{e}_{33}\overline{\lambda}_{33} + \overline{p}_3\overline{c}_{33}}{\overline{\varepsilon}_{33}\overline{c}_{33} + \overline{e}_{33}\overline{e}_{33}},
\tag{7.23}
$$

which shows that the pyroelectric coupling contributes to the carrier distribution directly and that the thermoelastic coupling does so through piezoelectric coupling. Then the following fields can be calculated:

$$S_{33} = -\frac{\overline{e}_{33}G_0\theta}{\overline{c}_{33}}\frac{\cosh kx_3}{\cosh kL} + \frac{\overline{\lambda}_{33}\theta}{\overline{c}_{33}},$$
$$E_3 = -G_0\theta\frac{\cosh kx_3}{\cosh kL}, \tag{7.24}$$

$$D_3 = G_1\left(1 - \frac{\cosh kz}{\cosh kL}\right)\theta,$$
$$P_3 = D_3 - \varepsilon_0 E_3 = G_1\theta\left[\left(\frac{\varepsilon_0}{\widehat{\varepsilon}_{33}} - 1\right)\frac{\cosh kx_3}{\cosh kL} + 1\right], \tag{7.25}$$

$$p = p_0\left(1 - \frac{q}{k_B\Theta_0}\frac{G_0\theta}{k}\frac{\sinh kx_3}{\cosh kL}\right),$$
$$n = n_0\left(1 + \frac{q}{k_B\Theta_0}\frac{G_0\theta}{k}\frac{\sinh kx_3}{\cosh kL}\right), \tag{7.26}$$

$$\rho^P = -P_{k,k} = -P_{3,3} = G_1\theta\left(1 - \frac{\varepsilon_0}{\widehat{\varepsilon}_{33}}\right)k\frac{\sinh kx_3}{\cosh kL}. \tag{7.27}$$

where

$$G_1 = \frac{\overline{e}_{33}\overline{\lambda}_{33}}{\overline{c}_{33}} + \overline{p}_3. \tag{7.28}$$

As a numerical example, consider a ZnO piezoelectric semiconductor rod. It is doped into an n-type semiconductor with $p \cong 0$. The length of the rod is $2L = 1.2$ μm. The cross-sectional area is 0.04 μm^2. Figure 7.2 shows the basic effect of interest, i.e., the temperature-induced carrier concentration perturbation. Under a temperature change, the mobile charges in the rod redistribute themselves under the electric field produced by the pyroelectric effect and the combination of

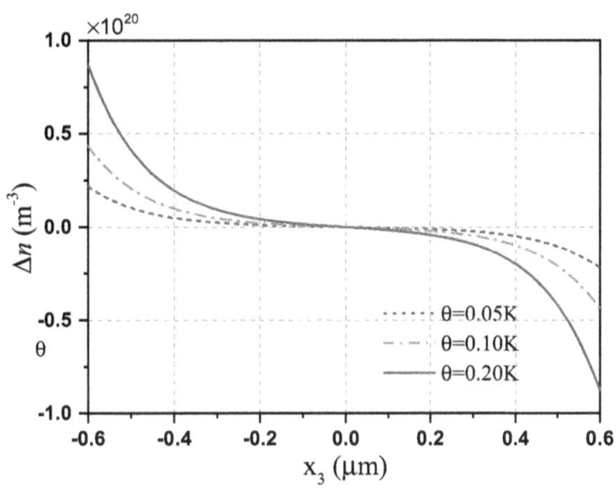

Fig. 7.2 Distribution of electron concentration perturbation along the rod for different temperature change. $n_0 = 10^{21}/\text{m}^3$

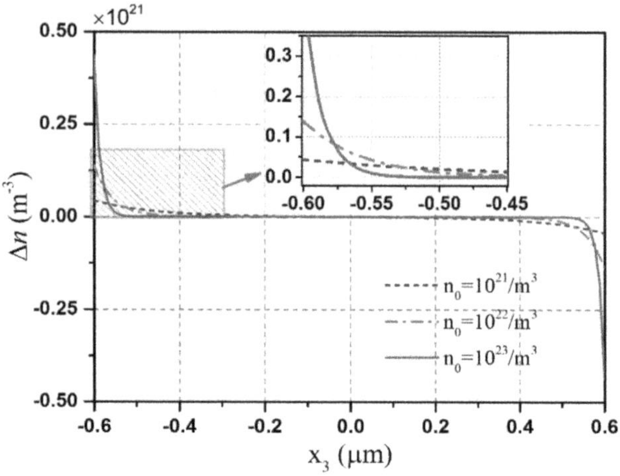

Fig. 7.3 Distribution of electron concentration perturbation along the rod for different n_0. θ=0.1 K

thermoelastic and piezoelectric effects. This thermally induced redistribution of charge carriers suggests the possibility of sensing or transduction between a temperature change and electric currents. In Fig. 7.2, Δn is an order of magnitude smaller than n_0 for the linearized theory to be valid.

Figure 7.3 shows Δn for different n_0 when θ=0.1 K, while all other parameters are kept the same as those for Fig. 7.2. Δn is sensitive to n_0.

7.2 Effects of a Local Temperature Change

From Sect. 3.8, it is known that local stresses in a piezoelectric semiconductor rod produce local electric potential barriers and wells and thus forbid the passage of currents under low voltages. We expect that a local temperature change in a piezoelectric semiconductor rod does the same and demonstrate it theoretically and numerically in this section [2]. Consider the ZnO rod as shown in Fig. 7.4. It is doped into an n-type semiconductor, and hence p is taken to be zero approximately. The rod is under a local and uniform temperature change within $|x_3| < a$. The problem is static.

The one-dimensional governing equations are taken from the previous section:

$$\begin{aligned} T_{33,3} &= \rho \ddot{u}_3, \\ D_{3,3} &= q(\Delta p - \Delta n), \end{aligned} \tag{7.29}$$

$$\begin{aligned} J^p_{3,3} &= -q\dot{p}, \\ J^n_{3,3} &= q\dot{n}, \end{aligned} \tag{7.30}$$

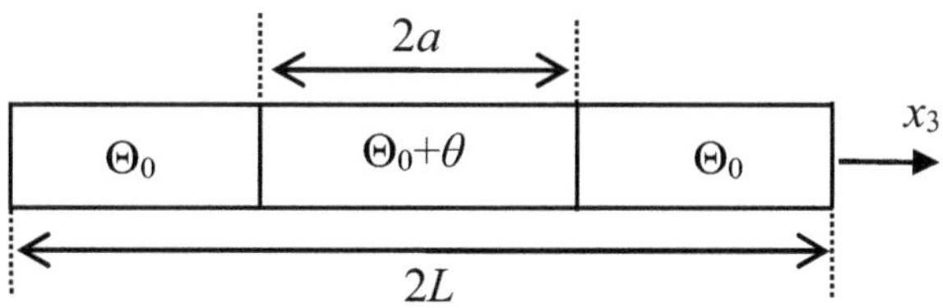

Fig. 7.4 A piezoelectric semiconductor rod under a local temperature change

$$\begin{aligned} T_{33} &= \overline{c}_{33}S_{33} - \overline{e}_{33}E_3 - \overline{\lambda}_{33}\theta, \\ D_3 &= \overline{e}_{33}S_{33} + \overline{\varepsilon}_{33}E_3 + \overline{p}_3\theta, \end{aligned} \tag{7.31}$$

$$\begin{aligned} J_3^p &= qp\mu_{33}^p E_3 - qD_3^p p_{,3} \\ &\cong qp_0\mu_{33}^p E_3 - qD_3^p(\Delta p)_{,3}, \\ J_3^n &= qn\mu_{33}^n E_3 + qD_3^n n_{,3}, \\ &\cong qn_0\mu_{33}^n E_3 + qD_3^n(\Delta n)_{,3}, \end{aligned} \tag{7.32}$$

$$S_{33} = u_{3,3}, \quad E_3 = -\varphi_{,3}. \tag{7.33}$$

The effective material constants in Eq. (7.31) are defined in Eq. (7.11). Successive substitutions of Eqs. (7.31), (7.32), and (7.33) into Eqs. (7.29) and (7.30) yield four second-order differential equations for u_3, φ, Δp, and Δn which are functions of x_3 and time. The equations may be linear or nonlinear depending on whether the linearized or nonlinear version of Eq. (7.32) is used.

We study the case when the rod is stress free and electrically open at its two ends first. We use the linearized constitutive relations for currents in Eq. (7.32) to obtain an analytical solution and then compare it with a numerical solution using the nonlinear version of Eq. (7.32). For simplicity, we denote the axial fields and material constants by

$$\begin{aligned} &x = x_3, \quad u = u_3, \\ &S = S_{33}, \quad T = T_{33}, \\ &E = E_3, \quad P = P_3, \quad D = D_3, \\ &J^p = J_3^p, \quad J^n = J_3^n, \end{aligned} \tag{7.34}$$

$$\begin{aligned} &\mu^p = \mu_{33}^p, \quad \mu^n = \mu_{33}^n, \\ &D^p = D_{33}^p, \quad D^n = D_{33}^n, \\ &c = \overline{c}_{33}, \quad e = \overline{e}_{33}, \quad \lambda = \overline{\lambda}_{33}, \\ &\overline{\varepsilon} = \overline{\varepsilon}_{33}, \quad \overline{p} = \overline{p}_3, \quad \alpha = \alpha_{33}. \end{aligned} \tag{7.35}$$

For the linear analytical analysis, we study the special case when $L = \infty$. The boundary and continuity conditions are

$$T(-\infty) = 0, \quad D(-\infty) = 0, \quad J^n(-\infty) = 0, \tag{7.36}$$

$$\begin{aligned}
&u(-a^-) = u(-a^+),\ \ T(-a^-) = T(-a^+),\\
&\varphi(-a^-) = \varphi(-a^+),\ \ D(-a^-) = D(-a^+),\\
&n(-a^-) = n(-a^+),\ \ J^n(-a^-) = J^n(-a^+),
\end{aligned} \tag{7.37}$$

$$\begin{aligned}
&u(a^-) = u(a^+),\ \ T(a^-) = T(a^+),\\
&\varphi(a^-) = \varphi(a^+),\ \ D(a^-) = D(a^+),\\
&n(a^-) = n(a^+),\ \ J^n(a^-) = J^n(a^+),
\end{aligned} \tag{7.38}$$

$$T(+\infty) = 0, \quad D(+\infty) = 0, \quad J^n(+\infty) = 0. \tag{7.39}$$

To determine the displacement and potential fields uniquely, we choose $-\infty$ as a reference and set

$$u(-\infty) = 0, \quad \varphi(-\infty) = 0. \tag{7.40}$$

The axial electric polarization P and the related distributed effective polarization charge density can be calculated from

$$\begin{aligned}
&P = D - \varepsilon_0 E,\\
&\rho^P = -P_{k,k} = -\frac{dP}{dx}.
\end{aligned} \tag{7.41}$$

Mathematically, we need to find solutions for each of the three regions in $x < -a$, $|x| < a$, and $x > a$, respectively, and apply boundary and continuity conditions. For the linear solution, in each region, we have a system of ordinary differential equations with constant coefficients. The final solution can be found in a straightforward manner. For $x < -a$, the results are

$$\begin{aligned}
u &= \frac{e}{c}\frac{(e\lambda + \overline{p}c)\theta}{k(e^2 + c\overline{\varepsilon})} \sinh ka \exp(kx),\\
S &= \frac{e}{c}\frac{(e\lambda + \overline{p}c)\theta}{e^2 + c\overline{\varepsilon}} \sinh ka \exp(kx),
\end{aligned} \tag{7.42}$$

$$\begin{aligned}
\varphi &= -\frac{(e\lambda + \overline{p}c)\theta}{k(e^2 + c\overline{\varepsilon})} \sinh ka \exp(kx),\\
E &= \frac{(e\lambda + \overline{p}c)\theta}{(e^2 + c\overline{\varepsilon})} \sinh ka \exp(kx),\\
D &= \left(\frac{e\lambda}{c} + \overline{p}\right)\theta \sinh ka \exp(kx),\\
P &= \left(\frac{e\lambda}{c} + \overline{p}\right)\theta \sinh ka\left(1 - \frac{c\varepsilon_0}{e^2 + c\overline{\varepsilon}}\right) \exp(kx),\\
\rho^P &= -k\left(\frac{e\lambda}{c} + \overline{p}\right)\theta \sinh ka\left(1 - \frac{c\varepsilon_0}{e^2 + c\overline{\varepsilon}}\right) \exp(kx),
\end{aligned} \tag{7.43}$$

$$\begin{aligned} \Delta n &= -\frac{k}{q}\left(\frac{e\lambda}{c}+\overline{p}\right)\theta \sinh ka \exp(kx), \\ \rho^P - q\Delta n &= k\left(\frac{e\lambda}{c}+\overline{p}\right)\theta \sinh ka \frac{c\varepsilon_0}{e^2+c\overline{\varepsilon}} \exp(kx), \end{aligned} \tag{7.44}$$

where

$$k^2 = \frac{n_0\mu^n}{D^n}\frac{q}{\varepsilon_{33}^T}. \tag{7.45}$$

In the central region with $|x| < a$, we have

$$\begin{aligned} u &= -\frac{e}{c}\frac{(e\lambda+\overline{p}c)\theta}{k(e^2+c\overline{\varepsilon})} \exp(-ka) \sinh kx + \frac{\lambda}{c}\theta(x+a), \\ S &= -\frac{e}{c}\frac{(e\lambda+\overline{p}c)\theta}{e^2+c\overline{\varepsilon}} \exp(-ka) \cosh kx + \frac{\lambda}{c}\theta, \end{aligned} \tag{7.46}$$

$$\begin{aligned} \varphi &= \frac{(e\lambda+\overline{p}c)\theta}{k(e^2+c\overline{\varepsilon})} \exp(-ka) \sinh kx, \\ E &= -\frac{(e\lambda+\overline{p}c)\theta}{e^2+c\overline{\varepsilon}} \exp(-ka) \cosh kx, \\ D &= -\left(\frac{e\lambda}{c}+\overline{p}\right)\theta \exp(-ka) \cosh kx + \left(\frac{e\lambda}{c}+\overline{p}\right)\theta, \\ P &= -\left(\frac{e\lambda}{c}+\overline{p}\right)\theta\left(1-\frac{c\varepsilon_0}{e^2+c\overline{\varepsilon}}\right) \exp(-ka) \cosh kx + \left(\frac{e\lambda}{c}+\overline{p}\right)\theta, \\ \rho^P &= k\left(\frac{e\lambda}{c}+\overline{p}\right)\theta\left(1-\frac{c\varepsilon_0}{e^2+c\overline{\varepsilon}}\right) \exp(-ka) \sinh kx, \end{aligned} \tag{7.47}$$

$$\begin{aligned} \Delta n &= \frac{k}{q}\left(\frac{e\lambda}{c}+\overline{p}\right)\theta \exp(-ka) \sinh kx, \\ \rho^P - q\Delta n &= -k\left(\frac{e\lambda}{c}+\overline{p}\right)\theta \frac{c\varepsilon_0}{e^2+c\overline{\varepsilon}} \exp(-ka) \sinh kx. \end{aligned} \tag{7.48}$$

For $x > a$, the fields are

$$\begin{aligned} u &= -\frac{e}{c}\frac{(e\lambda+\overline{p}c)\theta}{k(e^2+c\overline{\varepsilon})} \sinh ka \exp(-kx) + 2\frac{\lambda}{c}\theta L, \\ S &= \frac{e}{c}\frac{(e\lambda+\overline{p}c)\theta}{e^2+c\overline{\varepsilon}} \sinh ka \exp(-kx), \end{aligned} \tag{7.49}$$

$$\begin{aligned}
\varphi &= \frac{(e\lambda + \overline{p}c)\theta}{k(e^2 + c\overline{\varepsilon})} \sinh ka \exp(-kx), \\
E &= \frac{(e\lambda + \overline{p}c)\theta}{e^2 + c\overline{\varepsilon}} \sinh ka \exp(-kx), \\
D &= \left(\frac{e\lambda}{c} + \overline{p}\right)\theta \sinh ka \exp(-kx), \\
P &= \left(\frac{e\lambda}{c} + \overline{p}\right)\theta \sinh ka \left(1 - \frac{c\varepsilon_0}{e^2 + c\overline{\varepsilon}}\right) \exp(-kx), \\
\rho^P &= k\left(\frac{e\lambda}{c} + \overline{p}\right)\theta \sinh ka \left(1 - \frac{c\varepsilon_0}{e^2 + c\overline{\varepsilon}}\right) \exp(-kx),
\end{aligned} \tag{7.50}$$

$$\begin{aligned}
\Delta n &= \frac{k}{q}\left(\frac{e\lambda}{c} + \overline{p}\right)\theta \sinh ka \exp(-kx), \\
\rho^P - q\Delta n &= -k\left(\frac{e\lambda}{c} + \overline{p}\right)\theta \sinh ka \frac{c\varepsilon_0}{e^2 + c\overline{\varepsilon}} \exp(-kx).
\end{aligned} \tag{7.51}$$

For a numerical example, consider a ZnO rod with $2a = 1.2$ μm, $n_0 = 10^{21}/\text{m}^3$, and $\theta = 0.1$ K. As indicated by the above equations, a local temperature change produces many electromechanical fields. The situation is similar to those produced by local stresses. Our main interest is the local potential barrier and well shown in Fig. 7.5a which is caused by the thermally induced local polarization in (b). The polarization is under the screening effect of the electrons. As a linear solution, the potential barrier and well exhibit an antisymmetry about the origin.

The above potential barrier and well predicted by a linear analysis neglect the inherent nonlinearity in the drift currents. As a comparison, for the same rod with a finite length of $2L = 8.4$ μm, we perform a nonlinear analysis using COMSOL with larger values of θ. The results are shown in Fig. 7.6. In (a), when $\theta = 1$ K, the nonlinear result looks qualitatively similar to the linear result. However, when $\theta = 5$ K, fundamentally different from the linear result, the potential barrier height and well depth become clearly different, losing the antisymmetry of the linear solution about the origin. This is typical for a nonlinear solution. If θ is increased further, (b) shows that the well becomes very deep and the barrier is barely visible.

Next we consider the case when the two ends of a finite rod with length $2L$ is under an applied voltage $2V$ with the following boundary conditions:

$$\varphi(-L) = V, \quad \varphi(L) = -V. \tag{7.52}$$

Our main interest is the current-voltage relation or *I-V* curve. We use the nonlinear version of the constitutive relations for currents in Eq. (7.32) and perform a numerical analysis using COMSOL. Specifically, consider a rod with $2L = 8.4$ μm, $2a = 1.2$ μm, and $n_0 = 10^{21}/\text{m}^3$. The *I-V* curves of interest are shown in Fig. 7.7a for different θ, where $J^n = I$. For a specific curve corresponding to a given θ, when the applied voltage is low, no currents can flow through the rod in either direction. When the applied voltage is high and above a critical value that is temperature

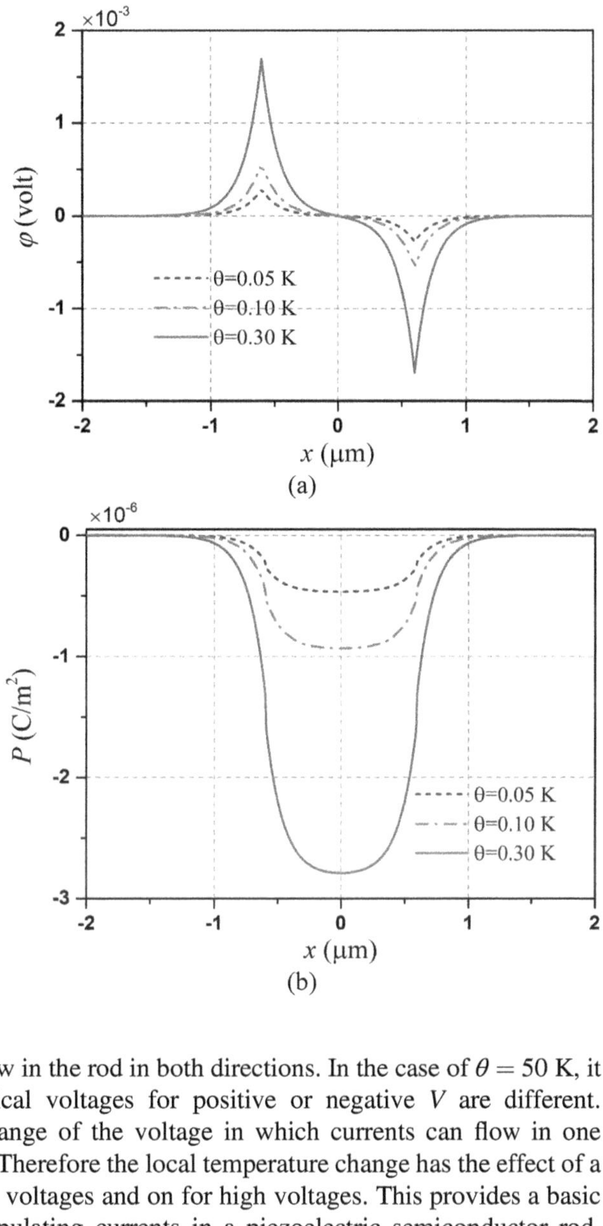

Fig. 7.5 Fields produced by a local temperature change predicted by the linear theory. (**a**) Potential barrier and well. (**b**) Electric polarization

dependent, currents can flow in the rod in both directions. In the case of $\theta = 50$ K, it can be seen that the critical voltages for positive or negative V are different. Therefore, there exists a range of the voltage in which currents can flow in one direction but not the other. Therefore the local temperature change has the effect of a switch which is off for low voltages and on for high voltages. This provides a basic means for thermally manipulating currents in a piezoelectric semiconductor rod. When $\theta = 35$ K, for specific values of the voltage that are low, critical, or high, (b) shows the corresponding distributions of the electric potential along the rod. When $V = 0.3$ volts, the potential well still exists, blocking possible currents. When $V = 1.5$ volts, it has overcome the potential well, and the potential distribution becomes monotonic. Hence currents can flow freely.

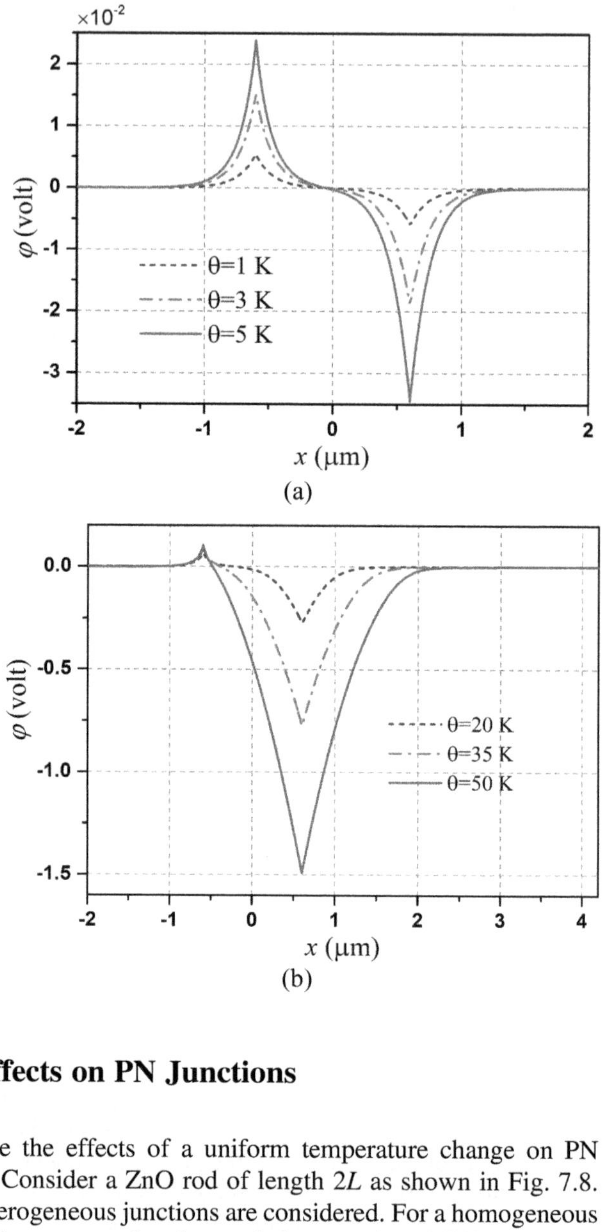

Fig. 7.6 Nonlinear solution for potential barrier and well by COMSOL. (**a**) $\theta = 1$, 3 and 5 K. (**b**) $\theta = 20$, 35, and 50 K

7.3 Temperature Effects on PN Junctions

In this section we examine the effects of a uniform temperature change on PN junctions in ZnO rods [3]. Consider a ZnO rod of length $2L$ as shown in Fig. 7.8. Both homogeneous and heterogeneous junctions are considered. For a homogeneous junction, the material is uniform except that the two halves are doped differently. For a heterogeneous junction, instead of using different materials, we simply reverse the c-axis of the left half. When the c-axis is reversed, the relevant piezoelectric and pyroelectric constants change signs. The temperature change θ is uniform. The problem is static. The rod is stress free and electrically open at both ends.

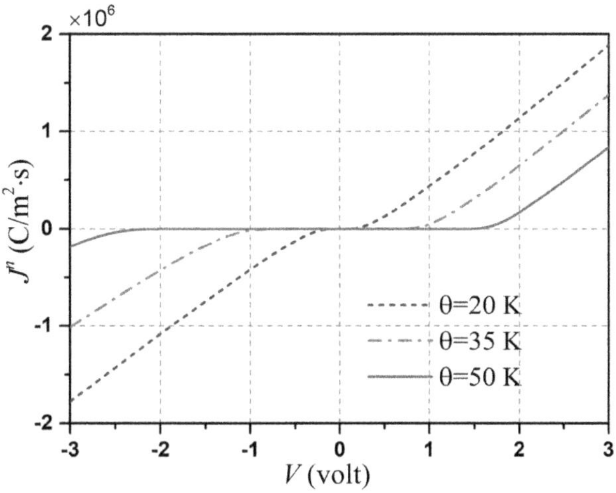

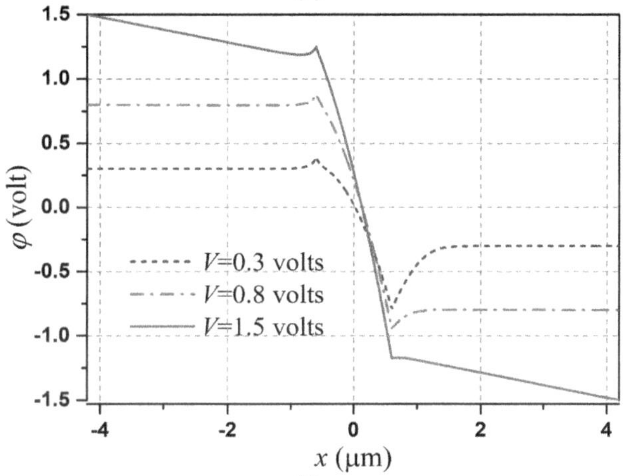

Fig. 7.7 Electrical behavior under a voltage. (**a**) *I-V* curves. (**b**) Potential distribution when $\theta = 35$ K

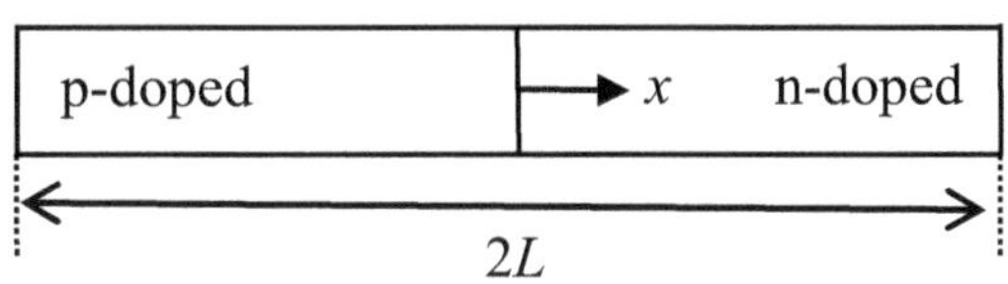

Fig. 7.8 A PN junction in a ZnO rod

The governing equations in Eqs. (7.29), (7.30), (7.31), (7.32), and (7.33) and the simplified notations in Eqs. (7.34) and (7.35) are employed. For an isolated rod, the boundary and continuity conditions are

$$\begin{aligned} T(-L) &= 0, \quad D(-L) = 0, \\ J^p(-L) &= 0, \quad J^n(-L) = 0, \end{aligned} \tag{7.53}$$

$$\begin{aligned} u(0^-) &= u(0^+), \quad T(0^-) = T(0^+), \\ \varphi(0^-) &= \varphi(0^+), \quad D(0^-) = D(0^+), \end{aligned} \tag{7.54}$$

$$\begin{aligned} p(0^-) &= p(0^+), \quad J^p(0^-) = J^p(0^+), \\ n(0^-) &= n(0^+), \quad J^n(0^-) = J^n(0^+), \end{aligned} \tag{7.55}$$

$$\begin{aligned} T(L) &= 0, \quad D(L) = 0, \\ J^p(L) &= 0, \quad J^n(L) = 0. \end{aligned} \tag{7.56}$$

Δp and Δn satisfy the following global charge neutrality conditions:

$$\int_{-L}^{L} \Delta p dx = 0, \quad \int_{-L}^{L} \Delta n dx = 0. \tag{7.57}$$

Only one of Eq. (7.57) is independent. To determine the displacement and potential fields uniquely, we choose a reference point, e.g., $x = a$, and set

$$u(a) = 0, \quad \varphi(a) = 0. \tag{7.58}$$

We need to find solutions in the two regions with $-L < x < 0$ and $0 < x < L$ separately and apply boundary and continuity conditions.

For the usual built-in fields in the junction, we perform a linear analysis analytically using the linearized constitutive relations in Eq. (7.32). In each region we have a system of ordinary differential equations with constant coefficients. The procedure for finding a general solution is straightforward. The junction may be heterogeneous. We use a prime for the material parameters in the left half and a double prime for those in the right half. The general solution for $-L < x < 0$ is

$$\Delta p - \Delta n = A_1 \sinh k'(x - L) + B_1 \sinh k'(x + L), \tag{7.59}$$

$$\varphi = -\frac{q}{(k')^2 \varepsilon'^T_{33}} [A_1 \sinh k'(x - L) + B_1 \sinh k'(x + L)] + C_1 x + C_2, \tag{7.60}$$

$$u = \frac{e'}{c'} \frac{q}{(k')^2 \varepsilon'^T_{33}} [A_1 \sinh k'(x - L) + B_1 \sinh k'(x + L)] + C_3 x + C_4, \tag{7.61}$$

$$\begin{aligned} \Delta p = \frac{p'_0 \mu'^p}{D'^p} \frac{q}{(k')^2 \varepsilon'^T_{33}} &[A_1 \sinh k'(x - L) + B_1 \sinh k'(x + L)] \\ &+ C_5 x + C_6, \end{aligned} \tag{7.62}$$

where A_1, B_1, and C_1-C_6 are undetermined constants and

$$(k')^2 = \left(\frac{p_0' \mu'^p}{D'^p} + \frac{n_0' \mu'^n}{D'^n} \right) \frac{q}{\varepsilon'^T_{33}}. \tag{7.63}$$

Similarly, for $0 < x < L$, we have

$$\Delta p - \Delta n = A_2 \sinh k''(x - L) + B_2 \sinh k''(x + L), \tag{7.64}$$

$$\varphi = -\frac{q}{(k'')^2 \varepsilon''^T_{33}} [A_2 \sinh k''(x - L) + B_2 \sinh k''(x + L)] + C_7 x + C_8, \tag{7.65}$$

$$u = \frac{e''}{c''} \frac{q}{(k'')^2 \varepsilon''^T_{33}} [A_2 \sinh k''(x - L) + B_2 \sinh k''(x + L)] + C_9 x + C_{10}, \tag{7.66}$$

$$\begin{aligned} \Delta p = \frac{p_0'' \mu''^p}{D''^p} \frac{q}{(k'')^2 \varepsilon''^T_{33}} [A_2 \sinh k''(x - L) + B_2 \sinh k''(x + L)] \\ + C_{11} x + C_{12}, \end{aligned} \tag{7.67}$$

where A_2, B_2, and C_7-C_{12} are undetermined constants and

$$(k'')^2 = \left(\frac{p_0'' \mu''^p}{D''^p} + \frac{n_0'' \mu''^n}{D''^n} \right) \frac{q}{\varepsilon''^T_{33}}. \tag{7.68}$$

Equations (7.59), (7.60), (7.61), and (7.62) and Eqs. (7.64), (7.65), (7.66), and (7.67) are substituted into Eqs. (7.53), (7.54), (7.55), (7.56), (7.57), and (7.58), resulting in a system of linear equations for the undetermined constants. These equations are solved on a computer. The axial electric polarization $P=P_3$ is calculated from Eq. (7.41). Then the related volume effective polarization charge density ρ^P, the sum of the doping and mobile charge densities ρ^e, and the total charge density ρ^t can be calculated from the following expressions:

$$\begin{aligned} \rho^P &= -P_{k,k} = -P_{3,3}, \\ \rho^e &= q(\Delta p - \Delta n), \\ \rho^t &= \rho^e + \rho^P. \end{aligned} \tag{7.69}$$

For a numerical example, consider a rod with $2L = 30$ μm. The doping profile is given by $p_0' = 10^{21}$ m^{-3}, $n_0' = 7 \times 10^{20}$ m^{-3}, $n_0'' = 10^{21}$ m^{-3}, and $p_0'' = 7 \times 10^{20}$ m^{-3}. It is chosen to be antisymmetric about the origin for simplicity. Consider three cases of temperature change with $\theta = 0$ K, 0.3 K, and 0.5 K, among which the case of $\theta = 0$ K is used as a reference.

Figure 7.9 shows the temperature effects on the basic built-in fields near the junction for a heterogeneous junction with opposite c-axes in the two halves. In the case of a homogeneous junction, the corresponding figures show no visible temperature effects and therefore are not presented. Only a portion of the rod within $|x| < 1$ μm is shown, which is sufficient for examining the fields because they are

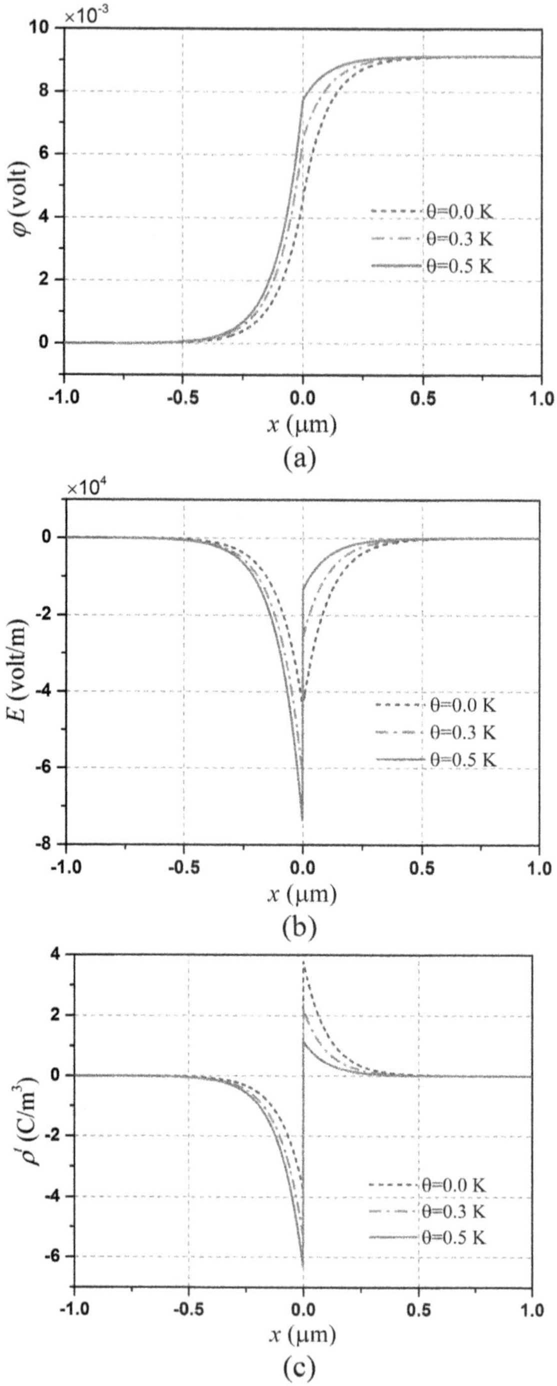

Fig. 7.9 Temperature effects on built-in fields in a heterogeneous junction. (**a**) Potential φ. (**b**) Electric field E. (**c**) Total charge density ρ^t

local and near $x = \pm 1$ μm they have already stabilized. Near the rod ends at far away with $x = \pm L$, there exist field concentrations which cannot be seen in the figure. When $\theta = 0$ K, the curves are either symmetric or antisymmetric about the origin. When θ is nonzero, the curves lose their symmetry or antisymmetry. Specifically, (a) and (b) show the usual built-in electric potential and field. When plotting the electric potential, $x = -1$ μm is the reference where the potential vanishes, i.e., $a = -1$ μm in Eq. (7.58). It can be seen that although the built-in voltage determined by the potential difference at $x = \pm 1$ μm in (a) is essentially unaffected by the temperature change, the slope of the potential field or the built-in electric field in (b) is. When $\theta = 0.5$ K, the potential in (a) is continuous but not smooth at the origin. This is related to that the corresponding electric field in (b) in fact has a jump discontinuity at the origin which has been connected by a vertical line in the figure. The total charge ρ^t in (c) is directly responsible for the behavior of the electric field in (b).

The difference between the homogenous and heterogeneous junctions is caused by the reversal of the c-axis in the left half of the rod, which leads to the fundamentally different polarizations shown in Fig. 7.10a, b for the homogeneous and heterogeneous junctions, respectively. Away from the junction and on its both sides, the P in (a) is the same but that in (b) has opposite signs. The different behaviors of P in (a) and (b) have implications on the effective polarization charge density ρ^P. For the homogeneous junction in (a), the corresponding ρ^P is essentially antisymmetric without any net charge and shows no visible temperature effect. Therefore it is not shown. For the heterogeneous junction in (b), the corresponding ρ^P is shown in (c) which produces some net positive charge near the junction. This affects the total charge density in Fig. 7.9c.

In addition to the volume effective polarization charge density described by ρ^P, there also exists surface effective polarization charge density σ^P on the interface of the PN junction due to the discontinuity of P there. For our junction interface, σ^P can be calculated from

$$\begin{aligned} \sigma^P &= P^- - P^+, \\ P^- &= P(0^-), \quad P^+ = P(0^+). \end{aligned} \tag{7.70}$$

We list σ^P for the same three temperature changes in Table 7.1. For a homogeneous junction or when $\theta = 0$ K, we have $\sigma^P = 0$. For a heterogeneous junction, σ^P increases with θ. σ^P is positive and contributes further to the net positive charge at the junction produced by ρ^P in Fig. 7.10c.

We now consider the case when the two ends of the rod at $x = \pm L$ are under an applied voltage $2V$ with the following boundary conditions:

$$\varphi(-L) = V, \quad \varphi(L) = -V. \tag{7.71}$$

Our main interest is the current-voltage relation or I-V curve. We use the nonlinear version of the constitutive relations for currents in Eq. (7.32) and perform a

Fig. 7.10 Polarization and its effective charges. (**a**) Polarization P, homogeneous junction. (**b**) Polarization P, heterogeneous junction. (**c**) ρ^P, heterogeneous junction

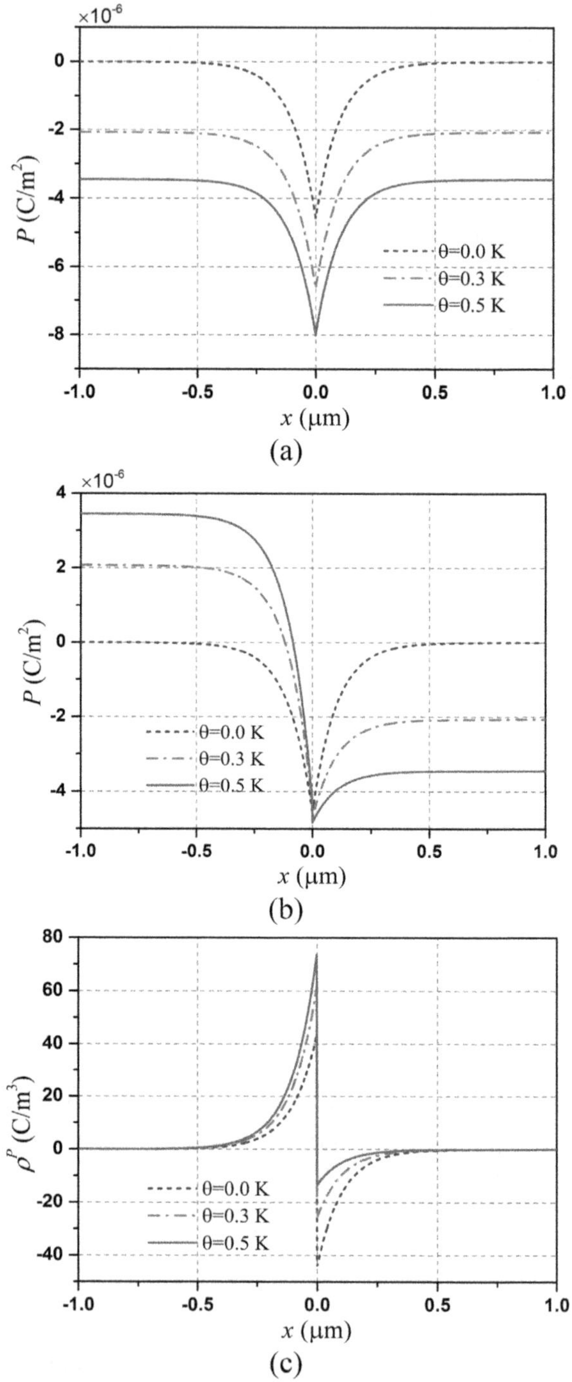

Table 7.1 Temperature effect on effective polarization charge density at junction interface

Charge (C/m^2)	P^-	P^+	σ^P
Temperature (K)	$\theta = 0$		
Homogeneous	-4.5638×10^{-6}	-4.5638×10^{-6}	0
Heterogeneous	-4.5638×10^{-6}	-4.5638×10^{-6}	0
Temperature (K)	$\theta = 0.3$		
Homogeneous	-6.6335×10^{-6}	-6.6335×10^{-6}	0
Heterogeneous	-4.4002×10^{-6}	-4.7274×10^{-6}	3.2724×10^{-7}
Temperature (K)	$\theta = 0.5$		
Homogeneous	-8.0134×10^{-6}	-8.0134×10^{-6}	0
Heterogeneous	-4.2911×10^{-6}	-4.8365×10^{-6}	5.4541×10^{-7}

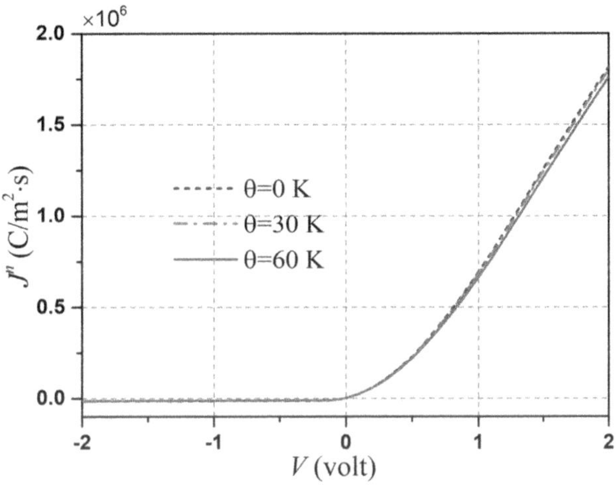

Fig. 7.11 *I-V* curves of the heterogeneous junction. $J^n = I$

numerical analysis using COMSOL. The *I-V* curves of the homogeneous junction are insensitive to the temperature change. The results for the heterogeneous junction are shown in Fig. 7.11.

From Sect. 7.1, it is known that the following combination of the effective one-dimensional material parameters

$$\overline{p} + \frac{e\lambda}{c} \tag{7.72}$$

plays a decisive role in the thermal behavior of a ZnO rod. It includes the effects of the effective one-dimensional pyroelectric, thermoelastic, and piezoelectric constants. With the use of Eq. (7.11), it can be shown that the combination in Eq. (7.72) is equal to the three-dimensional material constant p_3. For our ZnO rod, we have

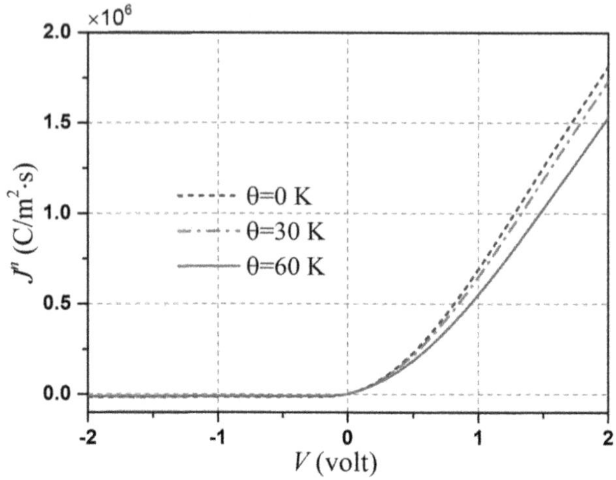

Fig. 7.12 *I-V* curves of the heterogeneous junction when the sign of *e* is artificially changed. $J^n = I$

$$\begin{aligned} \overline{p} &= -1.1981 \times 10^{-5} \ \left(\mathrm{C/m^2K}\right), \\ \frac{e\lambda}{c} &= 0.50808 \times 10^{-5} \ \left(\mathrm{C/m^2K}\right). \end{aligned} \tag{7.73}$$

Equation (7.73) shows that $\overline{p}$ and $e\lambda/c$ have opposite signs, and some cancelation is present in their combined effect. Therefore it can be expected that when $\overline{p}$ and e have the same sign, the temperature effect may be stronger. To verify this, we artificially change the sign of e in both halves of the junction but not that of $\overline{p}$ and plot the corresponding *I-V* curves in Fig. 7.12. Indeed, the temperature sensitivity is much stronger than that in Fig. 7.11.

7.4 Extension of Composite Rods

In this section we study the effects of a uniform temperature change on the extensional deformation of the composite rod shown in Fig. 7.13 [4]. It consists of two piezoelectric dielectric layers and a nonpiezoelectric semiconductor layer. (x,y,z) correspond to (x_1,x_2,x_3). When the piezoelectric materials are polarized ceramics or crystals of class (6mm), the poling direction or the c axis is along the z axis.

Equations (7.1), (7.2), (7.3), (7.4), and (7.5) are still valid. They include piezoelectric dielectrics and nonpiezoelectric semiconductors as special cases. We need to derive a set of one-dimensional equations from Eqs. (7.1), (7.2), (7.3), (7.4), and (7.5) for the extensional deformation of the composite rod under a uniform temperature change. For a thin composite rod, similar to Eqs. (7.6) and (7.7), we still have, for the entire rod, approximately

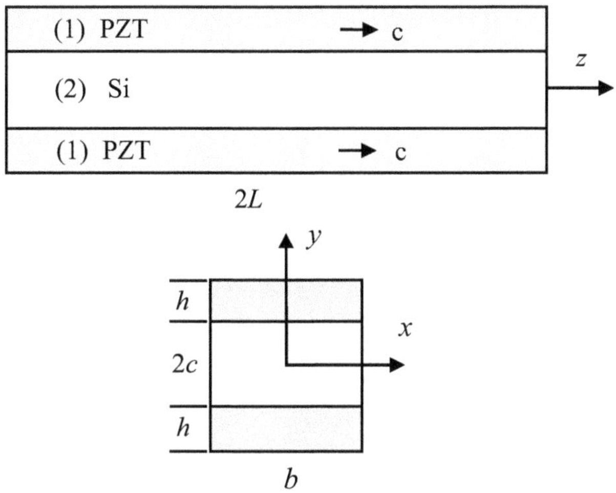

Fig. 7.13 Side view and cross section of a composite rod of piezoelectric dielectrics and nonpiezoelectric semiconductors

$$u_3 \cong u_3(z,t), \quad \varphi \cong \varphi(z,t). \tag{7.74}$$

In the semiconductor layer only, we also assume that

$$\Delta p \cong \Delta p(z,t), \quad \Delta n \cong \Delta n(z,t). \tag{7.75}$$

Then the axial strain and electric field can be approximated by

$$S_3 = S_{33} = u_{3,3}, \quad E_3 = -\varphi_{,3}, \tag{7.76}$$

which are common to all layers. To simplify the notation of the axial fields, we write

$$\begin{aligned} &u = u_3, \quad S = S_3, \quad T = T_{33} = T_3, \\ &E = E_3, \quad P = P_3, \quad D = D_3, \\ &J^p = J_3^p, \quad J^n = J_3^n. \end{aligned} \tag{7.77}$$

We also denote the relevant material constants by

$$\begin{aligned} &s = s_{3333}^E = s_{33}^E, \quad d = d_{333} = d_{33}, \quad \varepsilon = \varepsilon_{33}^T, \\ &\mu^p = \mu_{33}^p, \quad \mu^n = \mu_{33}^n, \quad D^p = D_{33}^p, \quad D^n = D_{33}^n, \\ &\alpha = \alpha_{33}. \end{aligned} \tag{7.78}$$

The fields and material constants for the piezoelectric layers will carry a superscript 1 in parentheses and those of the semiconductor layer a superscript 2 in parentheses.

For the piezoelectric layers, in terms of the elastic compliance, the constitutive relation for the axial strain and electric displacement are

$$\begin{aligned} S &= s^{(1)}T + d^{(1)}E + \alpha^{(1)}\theta, \\ D &= d^{(1)}T + \varepsilon^{(1)}E + p_3^{(1)}\theta. \end{aligned} \tag{7.79}$$

From Eq. (7.79) we obtain

$$\begin{aligned} T &= c^{(1)}S - e^{(1)}E - \lambda^{(1)}\theta, \\ D &= e^{(1)}S + \overline{\varepsilon}^{(1)}E + \overline{p}^{(1)}\theta, \end{aligned} \tag{7.80}$$

where

$$\begin{aligned} &c^{(1)} = \frac{1}{s^{(1)}}, \quad e^{(1)} = \frac{d^{(1)}}{s^{(1)}}, \quad \lambda^{(1)} = \frac{\alpha^{(1)}}{s^{(1)}}, \\ &\overline{\varepsilon}^{(1)} = \varepsilon^{(1)} - \frac{\left(d^{(1)}\right)^2}{s^{(1)}}, \quad \overline{p}^{(1)} = p_3^{(1)} - \frac{d^{(1)}\alpha^{(1)}}{s^{(1)}}. \end{aligned} \tag{7.81}$$

Similarly, for the semiconductor layer, it is assumed to be nonpiezoelectric and hence does not have pyroelectric coupling. The relevant constitutive relations are

$$\begin{aligned} S &= s^{(2)}T + \alpha^{(2)}\theta, \\ D &= \varepsilon^{(2)}E. \end{aligned} \tag{7.82}$$

Equation (7.82) can be rewritten into

$$\begin{aligned} T &= c^{(2)}S - \lambda^{(2)}\theta, \\ D &= \overline{\varepsilon}^{(2)}E, \end{aligned} \tag{7.83}$$

where

$$c^{(2)} = \frac{1}{s^{(2)}}, \quad \lambda^{(2)} = \frac{\alpha^{(2)}}{s^{(2)}}, \quad \overline{\varepsilon}^{(2)} = \varepsilon^{(2)}. \tag{7.84}$$

The currents exist in the semiconductor layer only and therefore the superscript 2 in parentheses is neglected:

$$\begin{aligned} J^p &\cong qp_0\mu^p E - qD^p\frac{d(\Delta p)}{dz}, \\ J^n &\cong qn_0\mu^n E + qD^n\frac{d(\Delta n)}{dz}. \end{aligned} \tag{7.85}$$

The total axial force and electric displacement in the composite rod are defined by integrations of T and D over the entire cross section and are found to be

$$\begin{aligned}\widehat{T} &= \widehat{c}S - \widehat{e}E - \widehat{\lambda}\theta,\\ \widehat{D} &= \widehat{e}S + \widehat{\varepsilon}E + \widehat{p}\theta,\end{aligned} \tag{7.86}$$

where

$$\begin{aligned}&\widehat{c} = c^{(1)}A^{(1)} + c^{(2)}A^{(2)}, \quad e = e^{(1)}A^{(1)},\\ &\widehat{\lambda} = \lambda^{(1)}A^{(1)} + \lambda^{(2)}A^{(2)},\\ &\widehat{\varepsilon} = \overline{\varepsilon}^{(1)}A^{(1)} + \overline{\varepsilon}^{(2)}A^{(2)}, \quad \widehat{p} = \overline{p}^{(1)}A^{(1)}.\end{aligned} \tag{7.87}$$

$A^{(1)}$ and $A^{(2)}$ are the cross-sectional areas of the piezoelectric and semiconductor layers, respectively. The one-dimensional equation of motion is obtained by applying Newton's second law to a differential element of the rod with length dz as shown in Fig. 7.14:

$$\frac{\partial \widehat{T}}{\partial z} + f(z,t) = 2b\left(\rho^{(1)}h + \rho^{(2)}c\right)\ddot{u}, \tag{7.88}$$

where $f(z,t)$ is the axial load per unit length of the rod. Similarly, the one-dimensional charge equation can be obtained by considering the differential element in Fig. 7.14 under electrical loads:

$$\frac{d\widehat{D}}{dz} = A^{(2)}q(\Delta p - \Delta n). \tag{7.89}$$

The one-dimensional continuity equations are simply

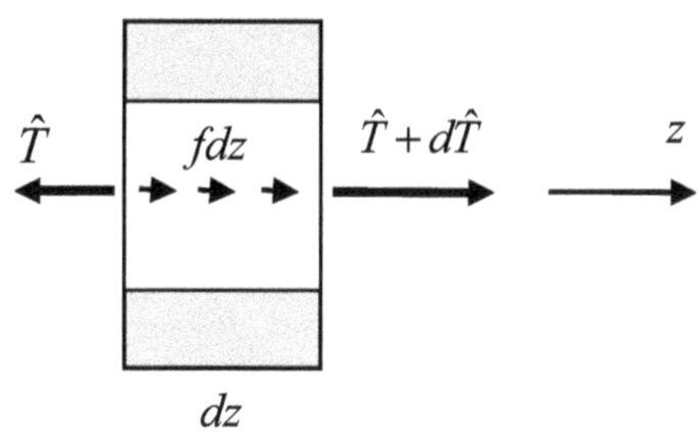

Fig. 7.14 A differential element of the rod

$$
\begin{aligned}
q\frac{\partial}{\partial t}(\Delta p) &= -\frac{\partial J^p}{\partial z}, \\
q\frac{\partial}{\partial t}(\Delta n) &= \frac{\partial J^n}{\partial z}.
\end{aligned} \tag{7.90}
$$

We consider the static extensional deformation of an electrically isolated free rod. The boundary conditions are

$$
\begin{aligned}
\widehat{T}(\pm L) = 0, \quad \widehat{D}(\pm L) = 0, \\
J^n(\pm L) = 0, \quad J^p(\pm L) = 0.
\end{aligned} \tag{7.91}
$$

Δp and Δn satisfy the following global charge neutrality conditions:

$$
\int_{-L}^{L} \Delta p dz = 0, \quad \int_{-L}^{L} \Delta n dz = 0. \tag{7.92}
$$

Only one of Eq. (7.92) is independent. To determine the mechanical displacement and the electric potential uniquely, we set

$$
u(0) = 0, \quad \varphi(0) = 0. \tag{7.93}
$$

Solving Eqs. (7.88), (7.89), and (7.90) under Eqs. (7.91), (7.92), and (7.93) does not present any mathematical challenge. Once the equations are solved, the polarization vector and effective polarization charge density can be calculated from their definitions:

$$
\begin{aligned}
P &= D - \varepsilon_0 E = \frac{\widehat{D}}{A^{(1)} + A^{(2)}} - \varepsilon_0 E, \\
\rho^P &= -P_{k,k} = -\frac{dP}{dz}.
\end{aligned} \tag{7.94}
$$

The expressions of the fields are found to be

$$
\varphi = \frac{\left(e\widehat{\lambda} + \widehat{pc}\right)\theta}{(\widehat{\varepsilon c} + e^2)k} \frac{\sinh kz}{\cosh kL}, \tag{7.95}
$$

$$
E = -\frac{\left(e\widehat{\lambda} + \widehat{pc}\right)\theta}{(\widehat{\varepsilon c} + e^2)} \frac{\cosh kz}{\cosh kL}, \tag{7.96}
$$

$$
u = -\frac{e\left(e\widehat{\lambda} + \widehat{pc}\right)\theta}{(\widehat{\varepsilon c} + e^2)\widehat{c}k} \frac{\sinh kz}{\cosh kL} + \frac{\widehat{\lambda}\theta}{\widehat{c}} z, \tag{7.97}
$$

$$S = -\frac{e\left(e\widehat{\lambda} + \widehat{p}\widehat{c}\right)\theta}{(\widehat{\varepsilon}\widehat{c} + e^2)\widehat{c}} \frac{\cosh kz}{\cosh kL} + \frac{\widehat{\lambda}\theta}{\widehat{c}}, \tag{7.98}$$

$$p = p_0 - \frac{p_0\mu^p}{D^p} \frac{\left(e\widehat{\lambda} + \widehat{p}\widehat{c}\right)\theta}{(\widehat{\varepsilon}\widehat{c} + e^2)k} \frac{\sinh kz}{\cosh kL}, \tag{7.99}$$

$$n = n_0 + \frac{n_0\mu^n}{D^n} \frac{\left(e\widehat{\lambda} + \widehat{p}\widehat{c}\right)\theta}{(\widehat{\varepsilon}\widehat{c} + e^2)k} \frac{\sinh kz}{\cosh kL}, \tag{7.100}$$

$$D = \left(\frac{\widehat{\lambda}e}{\widehat{c}} + \widehat{p}\right) \frac{\theta}{A^{(1)} + A^{(2)}} \left(1 - \frac{\cosh kz}{\cosh kL}\right), \tag{7.101}$$

$$\begin{aligned} P = &\left(\frac{\widehat{\lambda}e}{\widehat{c}} + \widehat{p}\right)\theta\left(\frac{\varepsilon_0}{\widetilde{\varepsilon}} \frac{1}{A^{(2)}} - \frac{1}{A^{(1)} + A^{(2)}}\right) \frac{\cosh kz}{\cosh kL} \\ &+ \left(\frac{\widehat{\lambda}e}{\widehat{c}} + \widehat{p}\right) \frac{\theta}{A^{(1)} + A^{(2)}}, \end{aligned} \tag{7.102}$$

$$\rho^P = \left(\frac{\widehat{\lambda}e}{\widehat{c}} + \widehat{p}\right)\left(\frac{1}{A^{(1)} + A^{(2)}} - \frac{\varepsilon_0}{\widetilde{\varepsilon}} \frac{1}{A^{(2)}}\right) k\theta \frac{\sinh \kappa z}{\cosh \kappa L}, \tag{7.103}$$

where

$$\begin{aligned} k^2 &= \left(\frac{p_0\mu^p}{D^p} + \frac{n_0\mu^n}{D^n}\right) \frac{q}{\widetilde{\widetilde{\varepsilon}}}, \\ \widetilde{\varepsilon}A^{(2)} &= \widehat{\varepsilon} + \frac{e^2}{\widehat{c}}. \end{aligned} \tag{7.104}$$

As a numerical example, consider a composite rod of PZT-5A and silicon with $L = 0.6$ μm, $h = c = 0.05$ μm, and $b = 0.2$ μm. $p_0 \cong 0$. $n_0 = 10^{23}/\text{m}^3$. Figure 7.15 shows the redistribution of electrons. It can be seen that the rod considered is sensitive to temperature and a larger temperature change induces larger changes of the carrier distribution as expected. Δn is an order of magnitude smaller than n_0 so that the linearized theory employed is valid.

Figure 7.16 shows the carrier concentration perturbation for different n_0 when θ=0.1 K while all other parameters are kept the same as those for Fig. 7.15. The fields are sensitive to n_0.

From Eq. (7.100), we obtain

$$\frac{\Delta n}{n_0} = \gamma \frac{\theta}{\Theta_0} \frac{\sinh kz}{\cosh kL}, \tag{7.105}$$

where

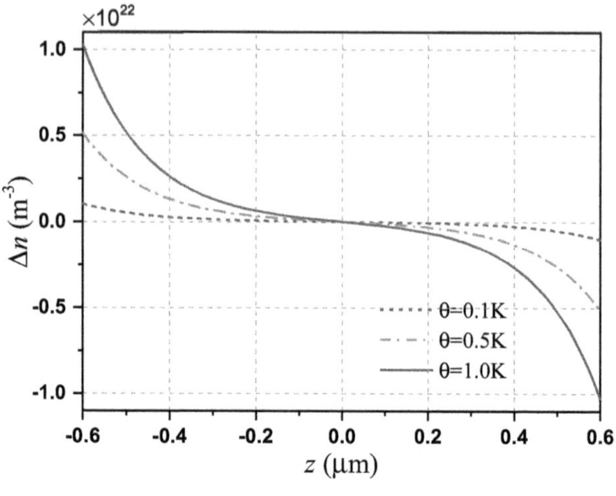

Fig. 7.15 Electron concentration perturbation in the rod for different temperature change. $n_0 = 10^{23}/\mathrm{m}^3$

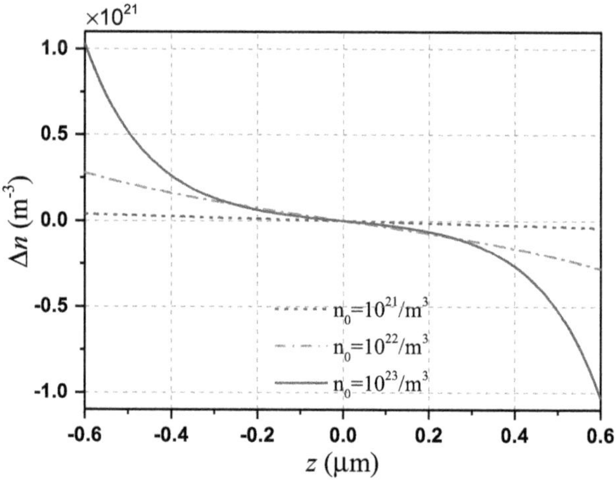

Fig. 7.16 Electron concentration perturbation in the rod for different n_0. θ=0.1 K

$$\begin{aligned} \gamma &= \frac{q}{k_B T} \frac{e}{\widetilde{\varepsilon}\widetilde{c}k} \frac{\widehat{\lambda} + \widehat{p}\widehat{c}}{A^{(1)} + A^{(2)}} \Theta_0, \\ \widetilde{c} &= \frac{\widehat{c} + e^2/\widehat{\varepsilon}}{A^{(1)} + A^{(2)}}. \end{aligned} \tag{7.106}$$

$\widetilde{c}$ is an averaged, piezoelectrically stiffened elastic constant. At the end of a long rod where kL is large, we have $\sinh kL \cong \cosh kL$. Therefore γ describes the relative change of carrier concentration near the ends of the rod per unit relative temperature change. It can be used as a measure of the strength of the coupling effect of the thermally induced carrier redistribution in the rod. We are interested in the change of number of carriers in the differential element in Fig. 7.14. From Eq. (7.105), for a

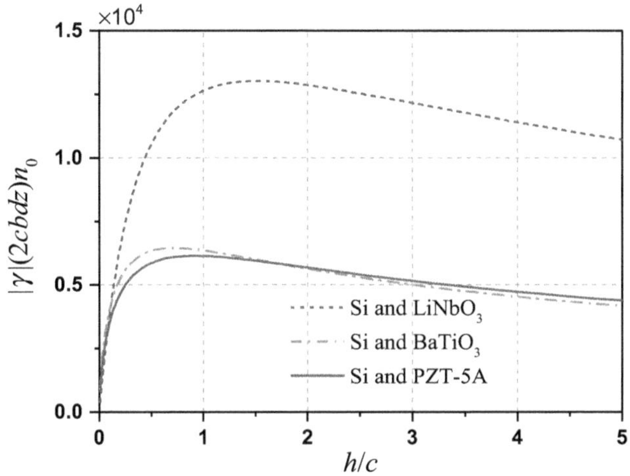

Fig. 7.17 $|\gamma|$ $(2cbdz)n_0$ at the end of the rod versus h/c for different combinations of materials. $dz = 0.1\mu\text{m}$. Θ_0=298 K. $p_0 \cong 0$. $n_0 = 10^{23}/\text{m}^3$

differential element of the rod near its ends, per unit θ/Θ_0, we have $\Delta n(2cbdz) \cong \gamma(2cbdz)n_0$ whose absolute value is plotted in Fig. 7.17 versus h/c when $h + c$ is fixed. The curves all have a maximum when h and c are not very different.

7.5 Extension and Bending of Composite Beams

In this section we study the behavior of the composite beam in Fig. 7.18 under a uniform temperature change [5]. It consists of a piezoelectric dielectric layer (layer (1)) such as ceramics poled along the axial direction and a nonpiezoelectric semiconductor layer (layer (2)) such as silicon. The cross-sectional areas of the piezoelectric and semiconductor layers are denoted by $A^{(1)}$ and $A^{(2)}$. The beam is unelectroded. The x_3 axis is at the interface which is in both extension and bending under a uniform temperature change because of the lack of structural symmetry about the interface.

Equations (7.1), (7.2), (7.3), (7.4), and (7.5) are still valid, with piezoelectric dielectrics and nonpiezoelectric semiconductors as special cases. We derive a set of one-dimensional equations first. For extension and bending without shear deformation in the x_3-x_2 plane, the bending and axial displacements are approximated by

$$u_2 \cong v(x_3, t), \quad u_3 \cong w(x_3, t) - x_2 v_{,3}. \tag{7.107}$$

The axial strain S_3 is expressed in terms of w and v through

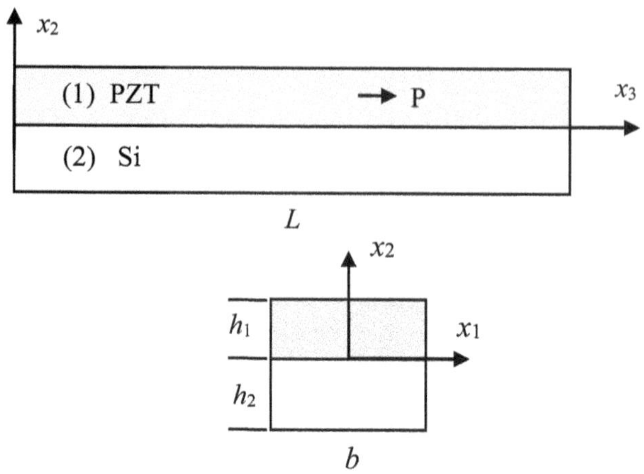

Fig. 7.18 Side view and cross section of a composite beam of a piezoelectric dielectric and a nonpiezoelectric semiconductor

$$S_3 = w_{,3} - x_2 v_{,33}. \tag{7.108}$$

Similarly, the electric potential and electric field over a cross section are approximated by

$$\varphi \cong \varphi(x_3, t), \quad E_3 = -\varphi_{,3}. \tag{7.109}$$

For the carrier concentration perturbations in the semiconductor layer,

$$\Delta p \cong \Delta p(x_3, t), \quad \Delta n \cong \Delta n(x_3, t). \tag{7.110}$$

We follow the simplified notations for the relevant axial fields and material parameters in Eqs. (7.77) and (7.78). For the piezoelectric layer, the one-dimensional constitutive relations and effective material constants in Eqs. (7.80) and (7.81) are still applicable. The constitutive relations take the following form with the use of Eqs. (7.108) and (7.109):

$$\begin{aligned} T &= c^{(1)}S - e^{(1)}E - \lambda^{(1)}\theta \\ &= c^{(1)}(w_{,3} - x_2 v_{,33}) + e^{(1)}\varphi_{,3} - \lambda^{(1)}\theta, \\ D &= e^{(1)}S + \overline{\varepsilon}^{(1)}E + \overline{p}^{(1)}\theta \\ &= e^{(1)}(w_{,3} - x_2 v_{,33}) - \overline{\varepsilon}^{(1)}\varphi_{,3} + \overline{p}^{(1)}\theta. \end{aligned} \tag{7.111}$$

Similarly, for the semiconductor layer, Eqs. (7.83) and (84) are still applicable. The one-dimensional constitutive relations take the following form with the use of Eqs. (7.108) and (109):

$$
\begin{aligned}
T &= c^{(2)}S - \lambda^{(2)}\theta \\
&= c^{(2)}(w_{,3} - x_2 v_{,33}) - \lambda^{(2)}\theta, \\
D &= \varepsilon^{(2)}E = -\varepsilon^{(2)}\varphi_{,3}.
\end{aligned} \tag{7.112}
$$

The currents exist in the semiconductor layer only and therefore the superscript 2 in parentheses is neglected. We still have Eq. (7.85):

$$
\begin{aligned}
J^p &\cong qp_0\mu^p E - qD^p \frac{d(\Delta p)}{dx_3}, \\
J^n &\cong qn_0\mu^n E + qD^n \frac{d(\Delta n)}{dx_3}.
\end{aligned} \tag{7.113}
$$

The total axial force $\widehat{T}$, shear force Q, bending moment M, and electric displacement $\widehat{D}$ are defined by the following integrations over a cross section:

$$
\begin{aligned}
\widehat{T} &= \int T dx_1 dx_2 = b\int_0^{h_1} T dx_2 + b\int_{-h_2}^{0} T dx_2 \\
&= \left(c^{(1)}w_{,3} + e^{(1)}\varphi_{,3} - \lambda^{(1)}\theta\right)h_1 b \\
&\quad - \frac{c^{(1)}v_{,33}}{2}h_1^2 b + \left(c^{(2)}w_{,3} - \lambda^{(2)}\theta\right)h_2 b + \frac{c^{(2)}v_{,33}}{2}h_2^2 b,
\end{aligned} \tag{7.114}
$$

$$
Q = \int T_{32} dx_1 dx_2, \tag{7.115}
$$

$$
\begin{aligned}
M &= \int T x_2 dx_1 dx_2 = b\int_0^{h_1} T x_2 dx_2 + b\int_{-h_2}^{0} T x_2 dx_2 \\
&= \frac{c^{(1)}w_{,3} + e^{(1)}\varphi_{,3} - \lambda^{(1)}\theta}{2}h_1^2 b \\
&\quad - \frac{c^{(1)}v_{,33}}{3}h_1^3 b + \frac{\lambda^{(2)}\theta - c^{(2)}w_{,3}}{2}h_2^2 b - \frac{c^{(2)}v_{,33}}{3}h_2^3 b,
\end{aligned} \tag{7.116}
$$

$$
\begin{aligned}
\widehat{D} &= \int D dx_1 dx_2 = b\int_0^{h_1} D dx_2 + b\int_{-h_2}^{0} D dx_2 \\
&= \left(e^{(1)}w_{,3} - \overline{\varepsilon}^{(1)}\varphi_{,3} + \overline{p}^{(1)}\theta\right)h_1 b \\
&\quad - \frac{e^{(1)}v_{,33}}{2}h_1^2 b + \left(-\varepsilon^{(2)}\varphi_{,3}\right)h_2 b.
\end{aligned} \tag{7.117}
$$

The constitutive relation for the shear force Q will be provided by the moment-shear force relationship as to be seen in the following. From the equations of motion (Newton's second law) of the differential element of the beam in Fig. 7.19 in the x_2 and x_3 directions as well as its moment equation, we have

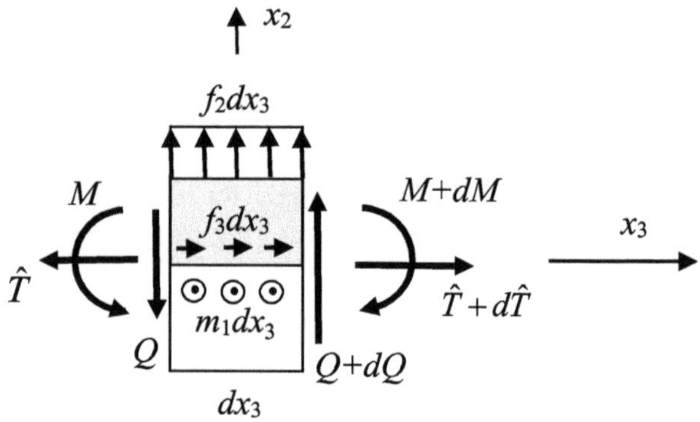

Fig. 7.19 A differential element of the beam under axial forces, shear forces, bending moments, and distributed mechanical loads

$$\begin{aligned}
&\widehat{T}_{,3} + f_3(x_3,t) = b\left(\rho^{(1)}h_1 + \rho^{(2)}h_2\right)\ddot{w},\\
&Q_{,3} + f_2(x_3,t) = b\left(\rho^{(1)}h_1 + \rho^{(2)}h_2\right)\ddot{v},\\
&M_{,3} - Q - m_1(x_3,t) = 0,
\end{aligned} \tag{7.118}$$

where $f_2(x_3,t)$ and $f_3(x_3,t)$ are the transverse and axial loads per unit length of the beam at $x_2 = 0$. $m_1(x_3,t)$ is the distributed moment per unit length of the beam. Equation (7.118)$_3$ is the moment-shear force relationship needed. Equations (7.118)$_3$ and (7.116) imply that

$$\begin{aligned}
Q = &\frac{c^{(1)}w_{,33} + e^{(1)}\varphi_{,33}}{2}h_1^2 b - \frac{c^{(1)}v_{,333}}{3}h_1^3 b\\
&-\frac{c^{(2)}w_{,33}}{2}h_2^2 b - \frac{c^{(2)}v_{,333}}{3}h_2^3 b + m_1,
\end{aligned} \tag{7.119}$$

which serves as the constitutive relation for the shear force. Similar to the derivation of Eq. (7.118), by considering the differential element in Fig. 7.19 under electrical loads, the charge equation of electrostatics can be written as

$$\widehat{D}_{,3} = q(\Delta p - \Delta n)A^{(2)}. \tag{7.120}$$

The one-dimensional conservation of holes and electrons are still given by Eq. (7.90):

$$\begin{aligned}
&q\frac{\partial}{\partial t}(\Delta p) = -J^p_{,3},\\
&q\frac{\partial}{\partial t}(\Delta n) = J^n_{,3}.
\end{aligned} \tag{7.121}$$

Consider static deformation of a free beam. There are no mechanical loads. $f_2 = 0$, $f_3 = 0$, and $m_1 = 0$. Equations (7.118), (7.120), and (7.121) reduce to

$$\begin{aligned} \widehat{T}_{,3} &= 0, \\ Q_{,3} &= M_{,33} = 0, \end{aligned} \tag{7.122}$$

$$\widehat{D}_{3,3} = q(\Delta p - \Delta n)A^{(2)}, \tag{7.123}$$

$$\begin{aligned} 0 &= -J^p_{3,3}, \\ 0 &= J^n_{3,3}. \end{aligned} \tag{7.124}$$

For the mechanically free and electrically isolated beam we are considering, the boundary conditions are

$$\begin{aligned} &\widehat{T}(0) = \widehat{T}(L) = 0, \quad M(0) = M(L) = 0, \\ &Q(0) = Q(L) = 0, \quad \widehat{D}(0) = \widehat{D}(L) = 0, \\ &J^p_3(0) = J^p_3(L) = 0, \quad J^n_3(0) = J^n_3(L) = 0. \end{aligned} \tag{7.125}$$

Δp and Δn must satisfy the following global charge neutrality conditions:

$$\int_0^L \Delta p dx_3 = 0, \quad \int_0^L \Delta n dx_3 = 0. \tag{7.126}$$

Only one of Eq. (7.126) is independent. To determine the mechanical displacements and the electric potential uniquely, we set

$$v(0) = 0, \quad v(L) = 0, \quad w(L/2) = 0, \tag{7.127}$$

$$\varphi(L/2) = 0. \tag{7.128}$$

Equations (7.122), (7.123), and (7.124) can be written as a fourth-order equation mainly for v, a second-order equation mainly for w, and second-order equations mainly for φ, Δp, and Δn, respectively, with couplings among them. The four equations can be further manipulated into

$$\begin{aligned} \varphi_{,33} - k^2\varphi &= -\frac{(C_1 + C_2)A^{(2)}}{\widetilde{\varepsilon}}, \\ v_{,333} &= \alpha\varphi_{,33}, \\ w_{,33} &= \beta\varphi_{,33}, \end{aligned} \tag{7.129}$$

$$
\begin{aligned}
q\Delta p &= -\frac{qp_0\mu_{33}^p}{D_{33}^p}\varphi + C_1, \\
-q\Delta n &= -\frac{qn_0\mu_{33}^n}{D_{33}^n}\varphi + C_2,
\end{aligned} \tag{7.130}
$$

where C_1 and C_2 are undetermined constants and

$$
\begin{aligned}
&A^{(1)} = h_1 b, \quad A^{(2)} = h_2 b, \\
&\alpha = \frac{6c^{(2)}e^{(1)}h_1h_2(h_1 + h_2)}{c^{(1)}c^{(1)}h_1^4 + c^{(2)}c^{(2)}h_2^4 + 2c^{(1)}c^{(2)}h_1h_2\left(2h_1^2 + 3h_1h_2 + 2h_2^2\right)}, \\
&\beta = -\frac{e^{(1)}h_1\left(c^{(1)}h_1^3 + c^{(2)}h_2^2(3h_1 + 4h_2)\right)}{c^{(1)}c^{(1)}h_1^4 + c^{(2)}c^{(2)}h_2^4 + 2c^{(1)}c^{(2)}h_1h_2\left(2h_1^2 + 3h_1h_2 + 2h_2^2\right)},
\end{aligned} \tag{7.131}
$$

$$
\begin{aligned}
\widetilde{\varepsilon} &= \overline{\varepsilon}^{(1)}A^{(1)} + \varepsilon^{(2)}A^{(2)} + \frac{e^{(1)}A^{(1)}h_1}{2}\alpha - e^{(1)}A^{(1)}\beta, \\
k^2 &= \frac{q}{\widetilde{\varepsilon}}\left(\frac{p_0\mu_{33}^p}{D_{33}^p} + \frac{n_0\mu_{33}^n}{D_{33}^n}\right)A^{(2)}.
\end{aligned} \tag{7.132}
$$

The general solution for φ, v, and w are found to be

$$
\begin{aligned}
\varphi &= kC_3 \sinh kx_3 + kC_4 \sinh k(x_3 - L) + \frac{1}{k^2\widetilde{\varepsilon}}(C_1 + C_2)A^{(2)}, \\
v &= \alpha(C_3 \cosh kx_3 + C_4 \cosh k(x_3 - L)) + C_5x_3^2 + C_6x_3 + C_7, \\
w &= \beta(kC_3 \sinh kx_3 + kC_4 \sinh k(x_3 - L)) + C_8x_3 + C_9,
\end{aligned} \tag{7.133}
$$

where C_3-C_9 are undetermined constants. Then Δp and Δn are determined from Eq. (7.130). The undetermined constants are governed by Eqs. (7.125), (7.126), (7.127), and (7.128).

As an example, consider PZT-5A for the piezoelectric layer and Si for the semiconductor layer with L = 0.6 μm, $h_1 = h_2$ = 25 nm, and b = 50 μm. $n_0 = p_0 = 10^{23}/\text{m}^3$. Θ_0 = 298 K. Some of these parameters may be varied later. We are particularly interested in the temperature-induced redistributions of holes and electrons. They are described by the Δp and Δn in Fig. 7.20.

Figure 7.21 shows Δp and Δn for different n_0 when $\theta = 0.5$ K while all other parameters are kept the same as those for Fig. 7.20. The fields are sensitive to p_0 and n_0.

Figure 7.22 shows the change of number of electrons in a differential element with length dz at the left end of the beam, i.e., $\Delta n(0)h_2bdz$ which may be used as a measure of the strength of the effect under consideration. $\Delta n(0)h_2bdz$ is plotted versus h_1/h_2 when $h_1 + h_2$ is fixed. The curves all have a maximum when h_1 and h_2 are not very different, which is reasonable.

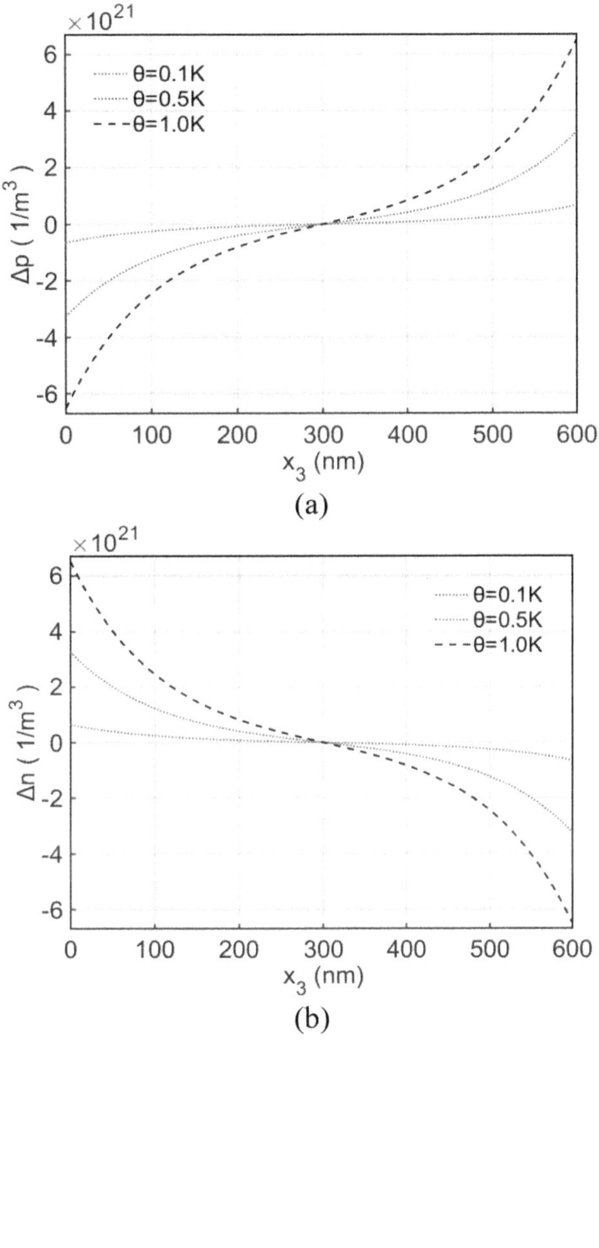

Fig. 7.20 (**a**) Hole concentration perturbation. (**b**) Electron concentration perturbation

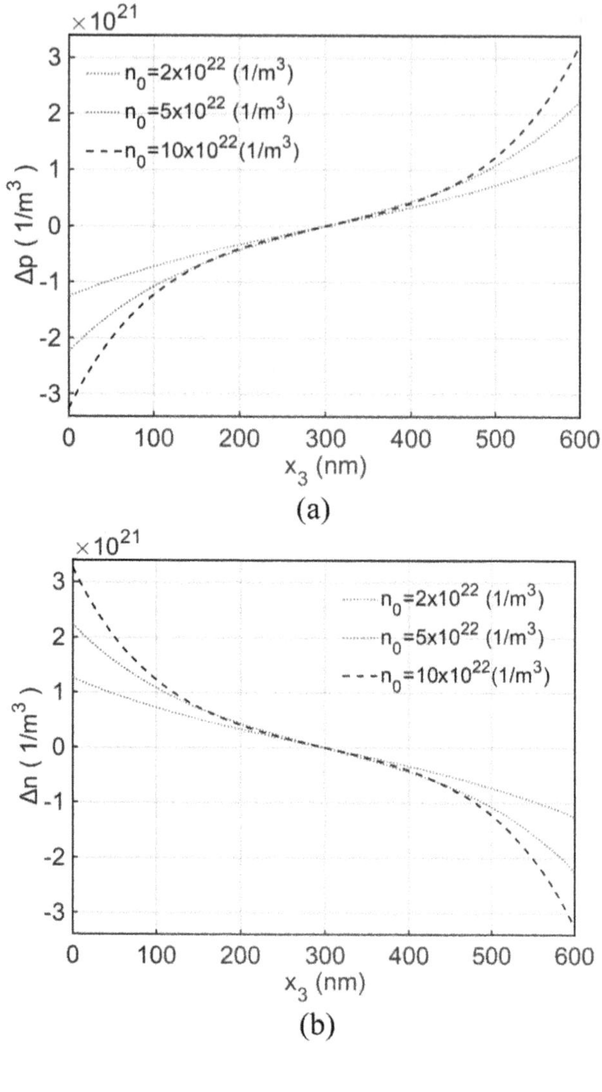

Fig. 7.21 Effects of $p_0 = n_0$. (**a**) Hole concentration perturbation. (**b**) Electron concentration perturbation

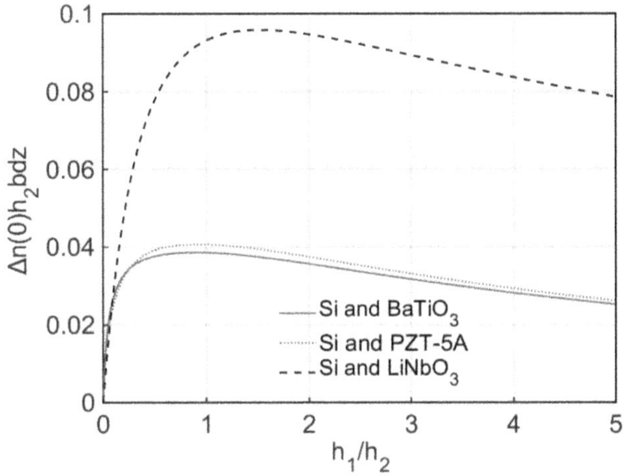

Fig. 7.22 $\Delta n(0)h_2bdz$ versus h_1/h_2 for different combinations of materials. $p_0 = n_0 = 10^{23}/\text{m}^3$, $\theta = 0.5$ K, $dz = 10$ nm

References

1. R.R. Cheng, C.L. Zhang, J.S. Yang, Thermally induced carrier distribution in a piezoelectric semiconductor fiber. J. Electron. Mater. **48**, 4939–4946 (2019)
2. R.R. Cheng, C.L. Zhang, W.Q. Chen, J.S. Yang, Electrical behaviors of a piezoelectric semiconductor fiber under a local temperature change. Nano Energy **66**, 104081 (2019)
3. R.R. Cheng, C.L. Zhang, W.Q. Chen, J.S. Yang, Temperature effects on PN junctions in piezoelectric semiconductor fibers with thermoelastic and pyroelectric couplings. J. Electron. Mater. **49**, 3140–3148 (2020)
4. R.R. Cheng, C.L. Zhang, W.Q. Chen, J.S. Yang, Temperature effects on mobile charges in extension of composite fibers of piezoelectric dielectrics and non-piezoelectric semiconductors. Int. J. Appl. Mech., **11**, 1950088 (2019)
5. Y.X. Luo, C.L. Zhang, W.Q. Chen, J.S. Yang, Thermally induced electromechanical fields in unimorphs of piezoelectric dielectrics and nonpiezoelectric semiconductors. Under review

Appendices

Appendix 1: List of Symbols

Because of the multiple fields involved and the use of one-, two-, and three-dimensional theories, some of the symbols have different meanings in different chapters or sections:

ε_0	Vacuum permittivity
μ_0	Vacuum permeability
q	Elementary charge
k_B	Boltzmann constant
i	Imaginary unit
$*$	Complex conjugate
i, j, k, l	Three-dimensional tensor indices. Range: 1, 2, 3
p, q	Three-dimensional matrix indices. Range: 1, 2, 3, 4, 5, 6
x_k	Cartesian coordinates
δ_{ij}	Kronecker delta
ρ	Mass density
f_j	Distributed body force
u_k	Mechanical displacement vector
T_{kl}	Stress tensor
S_{kl}	Strain tensor
φ	Electrostatic potential
ψ	Magnetic potential
E_i	Electric field
P_i	Electric polarization per unit volume
D_i	Electric displacement
B_i	Magnetic flux or induction
H_i	Magnetic field
p, n	Concentrations of holes and electrons
N_A^-, N_D^+	Concentrations of ionized accepters and donors

J. Yang, *Analysis of Piezoelectric Semiconductor Structures*,
https://doi.org/10.1007/978-3-030-48206-0

p_0, n_0	N_A^-, N_D^+
p^0, n^0	$p\|_{\varphi=0}$, $n\|_{\varphi=0}$
Δp, Δn	Carrier concentration perturbations
ρ^e	Sum of doping and mobile charge densities
ρ^t	Total charge density
Q^e	Charge
ρ^P	Volume effective polarization charge density
σ^P	Area effective polarization charge density
J_i^p, J_i^n	Current densities of holes and electrons
c_{ijkl}^E, s_{ijkl}^E	Elastic stiffness and compliance
e_{ijk}, d_{ijk}	Piezoelectric constants
ε_{ij}^S, ε_{ij}^T	Dielectric constants
μ_{ij}^p, μ_{ij}^n	Mobility of holes and electrons
D_{ij}^p, D_{ij}^n	Diffusion constants of holes and electrons
T	Absolute temperature
Θ_0	Reference temperature
θ	Temperature deviation from Θ_0
η	Entropy density
h_i	Heat flux
$\lambda_{\mathrm{i,j}}$	Thermoelastic constants
p_j	Pyroelectric constants
κ_{ij}	Heat conduction coefficients
D_{ij}^{Tp}, D_{ij}^{Tn}	Thermal diffusion constants
κ_{ij}^E, σ_{ij}^{Tp}, σ_{ij}^{Tn}	Thermoelectric constants
h_{ijk}	Piezomagnetic constants
α_{ij}	Magnetoelectric constants
μ_{ij}	Magnetic permeability
k_{15}^2, $\overline{k}_{15}^2$	Electromechanical coupling factor
N	Extensional force in rods
M	Bending moment in beams
Q	Shear force in beams; material quality factor
$\overline{c}_{33}$, $\overline{e}_{33}$, $\overline{\varepsilon}_{33}$	Effective material constants of rods/beams (Chaps. 3 and 4)
$\overline{c}_{pq}$, $\overline{e}_{ip}$, $\overline{\varepsilon}_{ij}$	Effective material constants of plates (Chap. 5)
$c_{pq}^{(0)}$, $e_{ip}^{(0)}$	Effective material constants of plates (Chap. 5)
γ_{rs}, ψ_{ks}, ζ_{kj}	Effective material constants of plates (Chap. 5)
κ, κ_1, κ_2	Thickness-shear correction factors of plates (Chap. 5)
a, b, c, d	Two-dimensional tensor indices of plates (Chaps. 5 and 6)
ω_∞, $\widehat{\omega}_\infty$	Fundamental thickness-shear frequency (Chaps. 5 and 6)
k	Wave number
λ_D	Debye-Hückel length

Appendix 2: Material Constants

Vacuum permittivity $\varepsilon_0 = 8.854 \times 10^{-12}$ F/m
Vacuum permeability $\mu_0 = 12.57 \times 10^{-7}$H/m
Elementary charge $q = 1.602 \times 10^{-19}$ C
Boltzmann constant k_B=1.381 × 10^{-23} J/K
$k_BT/q = 0.0259$V at room temperature 300 K [1]
Aluminum nitride (AlN) [2]

$$\rho = 3260\ \mathrm{kg/m^3}$$

$$[c_{pq}] = \begin{pmatrix} 345 & 125 & 120 & 0 & 0 & 0 \\ 125 & 345 & 120 & 0 & 0 & 0 \\ 120 & 120 & 395 & 0 & 0 & 0 \\ 0 & 0 & 0 & 118 & 0 & 0 \\ 0 & 0 & 0 & 0 & 118 & 0 \\ 0 & 0 & 0 & 0 & 0 & 110 \end{pmatrix} \times 10^9\ \mathrm{N/m^2}$$

$$[e_{ip}] = \begin{pmatrix} 0 & 0 & 0 & 0 & -0.48 & 0 \\ 0 & 0 & 0 & -0.48 & 0 & 0 \\ -0.58 & -0.58 & 1.55 & 0 & 0 & 0 \end{pmatrix} \mathrm{C/m^2}$$

$$[\varepsilon_{ij}] = \begin{pmatrix} 8.0 & 0 & 0 \\ 0 & 8.0 & 0 \\ 0 & 0 & 9.5 \end{pmatrix} \times 10^{-11}\ \mathrm{F/m}$$

Barium titanate ($BaTiO_3$) [3]

$$\rho = 5800\ \ \mathrm{kg/m^3}$$

$$[c_{pq}] = \begin{bmatrix} 166.0 & 77.0 & 78.0 & 0 & 0 & 0 \\ 77.0 & 166.0 & 78.0 & 0 & 0 & 0 \\ 78.0 & 78.0 & 162.0 & 0 & 0 & 0 \\ 0 & 0 & 0 & 43.0 & 0 & 0 \\ 0 & 0 & 0 & 0 & 43.0 & 0 \\ 0 & 0 & 0 & 0 & 0 & 44.5 \end{bmatrix} \times 10^9\ \ \mathrm{N/m^2}$$

$$[e_{jq}] = \begin{bmatrix} 0 & 0 & 0 & 0 & 11.6 & 0 \\ 0 & 0 & 0 & 11.6 & 0 & 0 \\ -4.4 & -4.4 & 18.6 & 0 & 0 & 0 \end{bmatrix} \mathrm{C/m^2}$$

$$[\varepsilon_{ij}] = \begin{bmatrix} 11.2 & 0 & 0 \\ 0 & 11.2 & 0 \\ 0 & 0 & 12.6 \end{bmatrix} \times 10^{-9} \ \text{F/m}$$

Cadium selenide (SdSe) [4]

$$\rho = 4820 \ \text{kg/m}^3$$

$$[c_{pq}] = \begin{bmatrix} 90.7 & 58.1 & 51.0 & 0 & 0 & 0 \\ 58.1 & 90.7 & 51.0 & 0 & 0 & 0 \\ 51.0 & 51.0 & 93.8 & 0 & 0 & 0 \\ 0 & 0 & 0 & 15.04 & 0 & 0 \\ 0 & 0 & 0 & 0 & 15.04 & 0 \\ 0 & 0 & 0 & 0 & 0 & 16.3 \end{bmatrix} \times 10^9 \ \text{N/m}^2$$

$$[e_{ip}] = \begin{bmatrix} 0 & 0 & 0 & 0 & -0.21 & 0 \\ 0 & 0 & 0 & -0.21 & 0 & 0 \\ -0.24 & -0.24 & 0.44 & 0 & 0 & 0 \end{bmatrix} \ \text{C/m}^2$$

$$[\varepsilon_{ij}] = \begin{bmatrix} 9.02 & 0 & 0 \\ 0 & 9.02 & 0 \\ 0 & 0 & 9.53 \end{bmatrix} \varepsilon_0$$

Germanium (Ge) [4, 5]

$$\rho = 5327 \ \text{kg/m}^3$$

$$[c_{pq}] = \begin{bmatrix} 128.9 & 48.3 & 48.3 & 0 & 0 & 0 \\ 48.3 & 128.9 & 48.3 & 0 & 0 & 0 \\ 48.3 & 48.3 & 128.9 & 0 & 0 & 0 \\ 0 & 0 & 0 & 67.1 & 0 & 0 \\ 0 & 0 & 0 & 0 & 67.1 & 0 \\ 0 & 0 & 0 & 0 & 0 & 67.1 \end{bmatrix} \times 10^9 \ \text{N/m}^2$$

$$[\varepsilon_{ij}] = \begin{bmatrix} 0.1398932 & 0 & 0 \\ 0 & 0.1398932 & 0 \\ 0 & 0 & 0.1398932 \end{bmatrix} \times 10^{-9} \ \text{F/m}$$

$$\mu_n = 3900 \ \text{cm}^2/\text{V} \cdot \text{s}$$

$$\mu_p = 1900 \ \text{cm}^2/\text{V} \cdot \text{s}$$

Lithium niobate ($LiNbO_3$) [6, 7]

$$\rho = 4700 \quad \text{kg/m}^3,$$

$$[c_{pq}] = \begin{pmatrix} 2.03 & 0.53 & 0.75 & 0.09 & 0 & 0 \\ 0.53 & 2.03 & 0.75 & -0.09 & 0 & 0 \\ 0.75 & 0.75 & 2.45 & 0 & 0 & 0 \\ 0.09 & -0.09 & 0 & 0.60 & 0 & 0 \\ 0 & 0 & 0 & 0 & 0.60 & 0.09 \\ 0 & 0 & 0 & 0 & 0.09 & 0.75 \end{pmatrix} \times 10^{11} \quad \text{N/m}^2$$

$$[e_{ip}] = \begin{pmatrix} 0 & 0 & 0 & 0 & 3.70 & -2.50 \\ -2.50 & 2.50 & 0 & 3.70 & 0 & 0 \\ 0.20 & 0.20 & 1.30 & 0 & 0 & 0 \end{pmatrix} \quad \text{C/m}^2$$

$$[\varepsilon_{ij}] = \begin{pmatrix} 38.9 & 0 & 0 \\ 0 & 38.9 & 0 \\ 0 & 0 & 25.7 \end{pmatrix} \times 10^{-11} \quad \text{F/m}$$

Lithium tantalate ($LiTaO_3$) [6, 7]

$$\rho = 7450 \quad \text{kg/m}^3,$$

$$[c_{pq}] = \begin{pmatrix} 2.33 & 0.47 & 0.80 & -0.11 & 0 & 0 \\ 0.47 & 2.33 & 0.80 & 0.11 & 0 & 0 \\ 0.80 & 0.80 & 2.75 & 0 & 0 & 0 \\ -0.11 & -0.11 & 0 & 0.94 & 0 & 0 \\ 0 & 0 & 0 & 0 & 0.94 & -0.11 \\ 0 & 0 & 0 & 0 & -0.11 & 0.93 \end{pmatrix} \times 10^{11} \quad \text{N/m}^2$$

$$[e_{ip}] = \begin{pmatrix} 0 & 0 & 0 & 0 & 2.6 & -1.6 \\ -1.6 & 1.6 & 0 & 2.6 & 0 & 0 \\ 0 & 0 & 1.9 & 0 & 0 & 0 \end{pmatrix} \quad \text{C/m}^2$$

$$[\varepsilon_{ij}] = \begin{pmatrix} 36.3 & 0 & 0 \\ 0 & 36.3 & 0 \\ 0 & 0 & 38.2 \end{pmatrix} \times 10^{-11} \quad \text{F/m}$$

PZT-2 [4]

$$\rho = 7600 \quad \text{kg/m}^3,$$

$$[c_{pq}] = \begin{pmatrix} 13.5 & 6.79 & 6.81 & 0 & 0 & 0 \\ 6.79 & 13.5 & 6.81 & 0 & 0 & 0 \\ 6.81 & 6.81 & 11.3 & 0 & 0 & 0 \\ 0 & 0 & 0 & 2.22 & 0 & 0 \\ 0 & 0 & 0 & 0 & 2.22 & 0 \\ 0 & 0 & 0 & 0 & 0 & 3.36 \end{pmatrix} \times 10^{10} \quad \text{N/m}^2$$

$$[e_{ip}] = \begin{pmatrix} 0 & 0 & 0 & 0 & 9.8 & 0 \\ 0 & 0 & 0 & 9.8 & 0 & 0 \\ -1.9 & -1.9 & 9.0 & 0 & 0 & 0 \end{pmatrix} \ \text{C/m}^2$$

$$[\varepsilon_{ij}] = \begin{pmatrix} 504\varepsilon_0 & 0 & 0 \\ 0 & 504\varepsilon_0 & 0 \\ 0 & 0 & 260\varepsilon_0 \end{pmatrix}$$

PZT-4 [3]

$$\rho = 7600 \quad \text{kg/m}^3$$

$$[c_{pq}] = \begin{bmatrix} 138.5 & 77.37 & 73.64 & 0 & 0 & 0 \\ 77.37 & 138.5 & 73.64 & 0 & 0 & 0 \\ 73.64 & 73.64 & 114.8 & 0 & 0 & 0 \\ 0 & 0 & 0 & 25.6 & 0 & 0 \\ 0 & 0 & 0 & 0 & 25.6 & 0 \\ 0 & 0 & 0 & 0 & 0 & 30.6 \end{bmatrix} \times 10^{9} \quad \text{N/m}^2$$

$$[e_{jq}] = \begin{bmatrix} 0 & 0 & 0 & 0 & 12.72 & 0 \\ 0 & 0 & 0 & 12.72 & 0 & 0 \\ -5.2 & -5.2 & 15.08 & 0 & 0 & 0 \end{bmatrix} \ \text{C/m}^2$$

$$[\varepsilon_{ij}] = \begin{bmatrix} 13.06 & 0 & 0 \\ 0 & 13.06 & 0 \\ 0 & 0 & 11.15 \end{bmatrix} \times 10^{-9} \quad \text{F/m}$$

PZT-5A [3]

$$\rho = 7750 \quad \text{kg/m}^3$$

$$[c_{pq}] = \begin{bmatrix} 99.201 & 54.016 & 50.778 & 0 & 0 & 0 \\ 54.016 & 99.201 & 50.778 & 0 & 0 & 0 \\ 50.788 & 50.788 & 86.856 & 0 & 0 & 0 \\ 0 & 0 & 0 & 21.1 & 0 & 0 \\ 0 & 0 & 0 & 0 & 21.1 & 0 \\ 0 & 0 & 0 & 0 & 0 & 22.6 \end{bmatrix} \times 10^9 \ \text{N/m}^2$$

$$[e_{jq}] = \begin{bmatrix} 0 & 0 & 0 & 0 & 12.322 & 0 \\ 0 & 0 & 0 & 12.322 & 0 & 0 \\ -7.209 & -7.209 & 15.118 & 0 & 0 & 0 \end{bmatrix} \ \text{C/m}^2$$

$$[\varepsilon_{ij}] = \begin{bmatrix} 15.3 & 0 & 0 \\ 0 & 15.3 & 0 \\ 0 & 0 & 15.0 \end{bmatrix} \times 10^{-9} \ \text{F/m}$$

PZT-5H [4]

$$\rho = 7500 \ \ \text{kg/m}^3$$

$$[c_{pq}] = \begin{pmatrix} 12.6 & 7.95 & 8.41 & 0 & 0 & 0 \\ 7.95 & 12.6 & 8.41 & 0 & 0 & 0 \\ 8.41 & 8.41 & 11.7 & 0 & 0 & 0 \\ 0 & 0 & 0 & 2.30 & 0 & 0 \\ 0 & 0 & 0 & 0 & 2.30 & 0 \\ 0 & 0 & 0 & 0 & 0 & 2.33 \end{pmatrix} \times 10^{10} \ \text{N/m}^2$$

$$[e_{ip}] = \begin{pmatrix} 0 & 0 & 0 & 0 & 17.0 & 0 \\ 0 & 0 & 0 & 17.0 & 0 & 0 \\ -6.5 & -6.5 & 23.3 & 0 & 0 & 0 \end{pmatrix} \ \text{C/m}^2$$

$$[\varepsilon_{ij}] = \begin{pmatrix} 1700\varepsilon_0 & 0 & 0 \\ 0 & 1700\varepsilon_0 & 0 \\ 0 & 0 & 1470\varepsilon_0 \end{pmatrix}$$

Silicon (Si) [4, 5]

$$\rho = 2332 \ \ \text{kg/m}^3$$

$$[c_{pq}] = \begin{pmatrix} 16.57 & 6.39 & 6.39 & 0 & 0 & 0 \\ 6.39 & 16.57 & 6.39 & 0 & 0 & 0 \\ 6.39 & 6.39 & 16.57 & 0 & 0 & 0 \\ 0 & 0 & 0 & 7.956 & 0 & 0 \\ 0 & 0 & 0 & 0 & 7.956 & 0 \\ 0 & 0 & 0 & 0 & 0 & 7.956 \end{pmatrix} \times 10^{10}\ \mathrm{N/m^2}$$

$$[\varepsilon_{ij}] = \begin{pmatrix} 11.7 & 0 & 0 \\ 0 & 11.7 & 0 \\ 0 & 0 & 11.7 \end{pmatrix} \varepsilon_0$$
$$= \begin{bmatrix} 0.1035918 & 0 & 0 \\ 0 & 0.1035918 & 0 \\ 0 & 0 & 0.1035918 \end{bmatrix} \times 10^{-9}\ \mathrm{F/m}$$

$$\mu_n = 1500\ \mathrm{cm^2/V\cdot s}$$
$$\mu_p = 450\ \mathrm{cm^2/V\cdot s}$$

Zinc oxide (ZnO) [4]

$$\rho = 5680\ \mathrm{kg/m^3}$$

$$[c_{pq}] = \begin{pmatrix} 20.97 & 12.11 & 10.51 & 0 & 0 & 0 \\ 12.11 & 20.97 & 10.51 & 0 & 0 & 0 \\ 10.51 & 10.51 & 21.09 & 0 & 0 & 0 \\ 0 & 0 & 0 & 4.247 & 0 & 0 \\ 0 & 0 & 0 & 0 & 4.247 & 0 \\ 0 & 0 & 0 & 0 & 0 & 4.43 \end{pmatrix} \times 10^{10}\ \mathrm{N/m^2}$$

$$[e_{ip}] = \begin{pmatrix} 0 & 0 & 0 & 0 & -0.48 & 0 \\ 0 & 0 & 0 & -0.48 & 0 & 0 \\ -0.573 & -0.573 & 1.32 & 0 & 0 & 0 \end{pmatrix} \mathrm{C/m^2}$$

$$[\varepsilon_{ij}] = \begin{pmatrix} 8.55 & 0 & 0 \\ 0 & 8.55 & 0 \\ 0 & 0 & 10.2 \end{pmatrix} \varepsilon_0$$

References

1. R.F. Pierret, *Semiconductor Device Fundamentals* (Addison-Wesley, Reading, 1996)
2. K. Tsubouchi, K. Sugai, N. Mikoshiba, AlN material constants evaluation and SAW properties on AlN/Al_2O_3 and AlN/Si, in *Proceedings of the IEEE Ultrasonics Symposium*, (1981), pp. 375–380
3. F. Ramirez, P.R. Heyliger, E. Pan, Free vibration response of two-dimensional magneto-electro-elastic laminated plates. J. Sound Vib. **292**, 626–644 (2006)
4. B.A. Auld, *Acoustic Fields and Waves in Solids*, vol 1 (Wiley, New York, 1973)
5. S.M. Sze, *Physics of Semiconductor Devices* (Wiley, New York, 1981)
6. A.W. Warner, M. Onoe, G.A. Couqin, Determination of elastic and piezoelectric constants for crystals in class (3m). J. Acoust. Soc. Am. **42**, 1223–1231 (1967)
7. H.F. Tiersten, *Linear Piezoelectric Plate Vibrations* (Plenum, New York, 1969)

Index

J. Yang, *Analysis of Piezoelectric Semiconductor Structures*,
https://doi.org/10.1007/978-3-030-48206-0

Ingram Content Group UK Ltd.
Milton Keynes UK
UKHW020204230523
422194UK00003B/106